HURRICANE KATRINA

PERFORMANCE OF TRANSPORTATION SYSTEMS

EDITED BY
Reginald DesRoches, Ph.D.

Technical Council on Lifeline Earthquake Engineering
Monograph No. 29
April 2006

Published by the American Socirty of Civil Engineers

Library of Congress Cataloging-in-Publication Data

Hurricane Katrina : performance of transportation systems / edited by Reginald DesRoches.
p. cm. -- (Technical council on lifeline earthquake engineering monograph ; no. 29)
Includes index.
ISBN 0-7844-0879-3
1. Transportation engineering. 2. Hurricane Katrina, 2005. I. DesRoches, Reginald.

TA1145.H87 2006
629.04--dc22

2006018865

Any statements expressed in these materials are those of the individual authors and do not necessarily represent the views of ASCE, which takes no responsibility for any statement made herein. No reference made in this publication to any specific method, product, process or service constitutes or implies an endorsement, recommendation, or warranty thereof by ASCE. The materials are for general information only and do not represent a standard of ASCE, nor are they intended as a reference in purchase specifications, contracts, regulations, statutes, or any other legal document.

ASCE makes no representation or warranty of any kind, whether express or implied, concerning the accuracy, completeness, suitability, or utility of any information, apparatus, product, or process discussed in this publication, and assumes no liability therefore. This information should not be used without first securing competent advice with respect to its suitability for any general or specific application. Anyone utilizing this information assumes all liability arising from such use, including but not limited to infringement of any patent or patents.

ASCE and American Society of Civil Engineers—Registered in U.S. Patent and Trademark Office.

Photocopies and reprints. You can obtain instant permission to photocopy ASCE publications by using ASCE's online permission service (www.pubs.asce.org/authors/RightslinkWelcomePage.html). Requests for 100 copies or more should be submitted to the Reprints Department, Publications Division, ASCE (address above); email: permissions@asce.org. A reprint order form can be found at www.pubs.asce.org/authors/reprints.html

0-7844-0879-3
Manufactured in the United States of America.

TCLEE Monograph Series

These publications may be purchased from ASCE, telephone 1-800-548-ASCE (2723), world wide web http://www.asce.org. The TCLEE publications web site is http://www.asce.org/instfound/techcomm_tclee.cfm Additional publications can be found at http://www.pubs.asce.org.

Kiremidjian, Ann, editor. *Recent Lifeline Seismic Risk Studies*. TCLEE Monograph No. 1, 1990.

Taylor, Craig, editor. *Seismic Loss Estimates for a Hypothetical Water System*. TCLEE Monograph No. 2, 1991.

Schiff, Anshel J., editor. *Guide to Post Earthquake Investigations of Lifelines*. TCLEE Monograph No. 3, 1991.

Cassaro, Michael, editor. *Lifeline Earthquake Engineering. Proceedings of the Third U.S. Conference*. TCLEE Monograph No. 4, August 1991.

Ballantyne, Donald, editor. *Lifeline Earthquake Engineering in the Central and Eastern U.S.* TCLEE Monograph No. 5, September 1992.

O'Rourke, Michael, editor. *Lifeline Earthquake Engineering. Proceedings of the Fourth U.S. Conference*. TCLEE Monograph No. 6, August 1995.

Schiff, Anshel J. and Buckle, Ian, editors. *Critical Issues and State of the Art on Lifeline Earthquake Engineering*. TCLEE Monograph No. 7, October 1995.

Schiff, Anshel J., editor. *Northridge Earthquake: Lifeline Performance and Post-Earthquake Response*. TCLEE Monograph No. 8, August 1995.

McDonough, Peter W., editor. *Seismic Design for Natural Gas Distributors*. TCLEE Monograph No. 9, August 1995.

Tang, Alex and Schiff, Anshel, editors. *Methods of Achieving Improved Seismic Performance of Communication Systems*. TCLEE Monograph No. 10, September 1996.

Schiff, Anshel, editor. *Guide to Post-Earthquake Investigation of Lifelines*. TCLEE Monograph No. 11, July 1997.

Werner, Stuart D., editor. *Seismic Guide to Ports*. TCLEE Monograph No. 12, March 1998.

Taylor, C, Mittler, E., Lund, L., editors. *Overcoming Barriers: Lifeline Seismic Improvement Programs*. TCLEE Monograph No. 13, September 1998.

Schiff, A.J., editor. *Hyogoken-Nanbu (Kobe) Earthquake of January 17, 1995—Lifeline Performance*. TCLEE Monograph No. 14, September 1998.

Eidinger, John M., and Avila, Ernesto A., editors. *Guidelines for the Seismic Upgrade of Water Transmission Facilities*. TCLEE Monograph No. 15, January 1999.

Elliott, William M. and McDonough, Peter, editors. *Optimizing Post Earthquake Lifeline System Reliability, Proceedings of the Fifth National Conference on Lifeline Earthquake Engineering*. TCLEE Monograph No. 16, September 1999.

Tang, Alex K., editor. *Izmit (Kocaeli) Earthquake of August 17, 1999 Including Duzce Earthquake of November 12, 1999—Lifeline Performance*. TCLEE Monograph No. 17, March 2000.

Schiff, Anshel J., and Tang, Alex K., editors. *Chi-Chi Taiwan Earthquake of September 21, 1999—Lifeline Performance*. TCLEE Monograph No. 18, 2000.

Eidenger, John M., editor. *Kujurat (Kutch) India M7.7Earthquake of January 26, 2001 and Napa M5.2 Earthquake of September 3, 2000*. TCLEE Monograph No. 19, 2001.

McDonough Peter, editor. *The Nisqually, Washington Earthquake of February 28, 2001—Lifeline Performance*. TCLEE Monograph No. 20, 2002.

Taylor, Craig E. and Van Marcke, Erik H., editors. *Acceptable Risk Process – Lifelines and Natural Hazards*, TCLEE Monograph No. 21, 2002.

Heubach, William F., editor. *Seismic Screening Checklists for Water and Wastewater Facilities*, TCLEE Monograph No. 22, 2003.

Edwards, Curtis L., editor. *Atico, Peru M_w 8.4 Earthquake of June 23, 2001 – Lifeline Performance*, TCLEE Monograph No. 23, 2003.

Lund, Le Val and Sepponen, Carl, editors. *Lifeline Performance El Salvador Earthquakes of January 13 and February 13, 2001*, TCLEE Monograph No. 24, 2003.

Beavers, James E., editor. *Advancing Mitigation Technologies and Disaster Response for Lifeline Systems*, TCLEE Monograph No. 25, 2003.
Scawthorn, Charles, Eidinger, John M., and Schiff, Anshel J., editors. *Fire Following Earthquake*, TCLEE Monograph No. 26, 2005.

Edwards, Curtis L., editor. *Zemmouri, Algeria, Mw 6.8 Earthquake of May 31, 2003*, TCLEE No. 27, 2004.

Mark Yashinsky, Mark, editor. *San Simeon Earthquake of December 22, 2003 and Denali, Alaska, Earthquake of November 3, 2002,* TCLEE Monograph No. 28, 2004.

DesRoches, Reginald, editor. *Hurricane Katrina: Performance of Transportation Systems*, TCLEE No. 29, 2006.

TCLEE Publications

Duke, C. Martin, editor. *The Current State of Knowledge of Lifeline Earthquake Engineering*, 1977.

Dowd, Munson, editor. *Annotated Bibliography on Lifeline Earthquake Engineering*, 1980.

Smith, D. J. Jr., editor. *Lifeline Earthquake Engineering: The Current State of Knowledge*, 1981.

Hall, William, editor. *Advisory Notes on Lifeline Earthquake Engineering,* 1983.

Nyman, Douglas, editor. *Guidelines for the Seismic Design of Oil and Gas Pipeline Systems, TCLEE Committee on Gas and Liquid Fuel Lifelines*, 1984.

Coopers, James, editor. *Lifeline Earthquake Engineering: Performance, Design and Construction*, 1984.

Eguchi, Ron, and Crouse, C. B., editors. *Lifeline Seismic Risk Analysis—Case Studies,* 1986.

Cassaro, Michael and Martinez-Romero, E., editors. *The Mexico Earthquakes—1985 Factors Involved and Lessons Learned,* 1986.

Wang, Leon R. L. and Whitman, Robert, editors. *Seismic Evaluation of Lifeline Systems—Case Studies*, 1986.

Cassaro, Michael and Cooper, James, editors. *Seismic Design and Construction of Complex Civil Engineering Systems*, 1988.

Werner, Stuart D. and Dickenson, Stephen E., editors. *Hyogoken-Nanbu (Kobe) Earthquake of January 17, 1995: A Post-Earthquake Reconnaissance of Port Facilities,* TCLEE Committee on Ports and Harbors Lifelines, 1996.

Schiff, Anshel J., editor. *Guide to Improved Earthquake Performance of Electric Power Systems*, ASCE Manual 96, 1999.

TCLEE Lifeline Earthquake Investigation Reports

ASCE Technical Council on Lifeline Engineering (TCLEE) Earthquake Investigation Committee (EIC) members have participated in a number of lifeline earthquake investigations. Reports have been prepared on the performance of lifeline and published in ASCE publications (other than TCLEE monographs), Earthquake Engineering Research Institute (EERI) and other publications listed as follows:

Coalinga Earthquake, May 2, 1983, M 6.5, Report by Luis Escalnate, TCLEE-EIC, EERI Newsletter.

Cluster County (Idaho) Earthquake, October 28, 1983, M 6.9, Report by Patrick Wong, TCLEE-EIC, ASCE News and EERI Newsletter.

Kaoiki (Hawaii) Earthquake, November 16, 1983, M 6.6, Report by Patrick Wong, TCLEE-EIC, ASCE News and EERI Newsletter.

Morgan Hill (California) Earthquake, April 24, 1984, M 6.1, Reports by Patrick Wong, TCLEE-EIC and Anshel Schiff, TCLEE-EIC, ASCE News.

Morgan Hill (California) Earthquake, April 24, 1984, EERI Earthquake Spectra, Vol. 1, No. 3, 1985.

Chile Earthquake, March 3, 1985, M 8.1, The Chile Earthquake of March 3, 1985, L. Escalante and A.J. Schiff, EERI Earthquake Spectra, Vol. 2, February 1986.

Mexico City Earthquake, September 19, 1985, M 8.1, Report by Luis Escalante, TCLEE, ASCE News, December 1985.

Mt. Lewis (Idaho) Earthquake, March 31, 1986, M 5.4, Report by Anshel Schiff, TCLEE-EIC, ASCE News, December 1986.

San Salvador Earthquake, October 10, 1986, M 5.4, Report by James Morgan, TCLEE-EIC, ASCE News, April-May 1986.

Ecuador Earthquakes, March 5, 1987, M 6.1 & 6.9, Report by Tom O'Rourke, ASCE News, June 1987.

Whittier Narrows Earthquake, October 1, 1987, M 5.9, The Whittier Narrows Earthquake of October 1, 1987, Responses of Lifelines and Effects on Emergency Response, A. J. Schiff, EERI Earthquake Spectra, Vol. 4, No. 2, February 1988.

Superstition Hills Earthquake, November 23, & 24, 1987, M 6.2 & 6.6, Report by Anshel Schiff, TCLEE-EIC, ASCE News, March 1988.

Pasadena Earthquake, December 3, 1988, M 4.9, Report by Le Val Lund, TCLEE-EIC, ASCE News.

Armenian Earthquake, December 7, 1988, M 6.9, Performance of Lifeline Systems, A. J. Schiff, ASCE-EIC, and S. Swan, EERI Earthquake Spectra, Armenian Earthquake Reconnaissance Report.

Loma Prieta Earthquake, October 17, 1989, M 7.1, Loma Prieta Earthquake Reconnaissance Report – Lifelines, Anshel Schiff and Le Val Lund, Coordinators, EERI Earthquake Spectra, Supplement to Vol. 6, May 1990.

Tejon Ranch Earthquake, Lifeline Response to the Tejon Ranch Earthquake, A. J. Schiff, EERI Earthquake Spectra, Vol. 5, No. 4, November 1989.

Philippine Earthquake, Philippine Earthquake Reconnaissance Report, A. J. Schiff, EERI Spectra Supplement to Vol. 7, 1991.

Sierra Madre Earthquake, June 28, 1991, M 5.8, Report by Le Val Lund, TCLEE-EIC, ASCE News, October 1991, EERI Special Earthquake Report, 1991; and Dames and Moore Earthquake Engineering News, Summer/Fall 1991.

Costa Rica Earthquake, Costa Rica Earthquake Reconnaissance Report, D. Ballantyne, EERI Earthquake Spectra, Supplement B to Vol. 7, October 1991.

Special Report, August 1989, Performance of Lifelines in the Spitak-88 Earthquake, A. J. Schiff, Proceedings of International Seminar on Spitak-88 Earthquake, Yerevan, Aremenia, May 23-26, 1989, UNESCO, 1992.

Erzincan, Turkey, Earthquake, Erzincan Turkey Reconnaissance Report, D. Ballantyne, EERI Spectra Supplement to Vol. 9, July 1993.

Landers and Big Bear Earthquakes, Lifelines Performance in the Landers and Big Bear (California) Earthquakes of 28 June 1992, Ms 7.5 and 6.6, Le Val Lund, Bulletin of the Seismological Society of America, Vol. 84, No. 3, June 1994.

Guam Earthquake, August 8, 1993, Ms 8.1, Guam Earthquake Reconnaissance Report, Craig D. Comartin, Technical Editor, EERI Supplement B to Volume II, April 1995.

Northridge Earthquake, Lifeline Utilities Performance on the 17 January 1994 Northridge, California Earthquake, Mw 6.7, Le Val Lund, Bulletin of the Seismological Society of American, Vol. 86, No. 1B, February 1996.

Hector Mine Earthquake, Lifeline Performance Hector Mine, CA., October 16, 1999 Earthquake, A Preliminary Reconnaissance Survey, Mw 7.1, Le Val Lund, American Society of Civil Engineers website (www.asce.org), October 2000.

Kocaeli, Turkey, Earthquake, Lifelines Performance, Kocaeli, Turkey, Earthquake, August 17, 1999, Mw 7.4, Le Val Lund, Business and Industry Council for Emergency Planning and Preparedness Bulletin, Winter 2000.

El Salvador Earthquakes, Preliminary Observations on the El Salvador Earthquakes of January 13 and February 13, 2001, M 7.6 and M 6.6, Le Val Lund et al., EERI Special Earthquake Report, July 2001.

Long Beach Earthquake, Lifelines Performance Long Beach Earthquake, March 10, 1933, A Historical Perspective, M 6.3, Le Val Lund, Business and Industry Council for Emergency Planning and Preparedness Bulletin, Summer 2002.

Denali, Alaska Earthquake, November 3, 2002, Mw 7.9, “Shaken”, Civil Engineering, Volume 73, Number 3, American Society of Civil Engineers, March 2003.

Zemmouri, Algeria Earthquake, Lifeline Damage Light to Moderate in Zemmouri Earthquake, Mw 6.6, Report by Curtis Edwards, Mark Yashinsky and Yumei Wang, TCLEE-EIC, ASCE News, August 2003.

Contents

HURRICANE KATRINA: PERFORMANCE OF TRANSPORTATION SYSTEMS

1 Executive Summary

Hurricane Katrina struck the gulf coast on August 29th, 2005, as an extremely large Category 3 storm. Wind speeds in excess of 140 MPH were recorded in parts of the gulf coast, and storm surges ranged from 25-35 feet in parts of Mississippi and Louisiana. While the failure of the levees in New Orleans has attracted most of the attention, there was significant damage to other infrastructure, including bridges, roads, and railroads. Approximately 45 bridges sustained damage, with the damage ranging from minor damage to railings or approach spans to major damage, as was seen in the collapse of several bridges in Mississippi and Louisiana. The I-10 twin span bridge across Lake Pontchartrain was the most catastrophic of the bridge failures, with over 473 spans shifting off their supports, and 64 spans completely fallen into the lake.

Rail operations were significantly disrupted due to either collapse of track carrying bridges, or debris such as trees and barges covering and damaging railroad tracks. However, in general, railroad bridges performed better than highway bridges during the hurricane. The major impedance to roadway traffic following hurricane Katrina was the extensive amount of debris from the storm. Debris removal from roads in the gulf coast region is estimated at approximately $200 million. The damage to bridges from Katrina will have a major impact on future bridge design practices in the region. Many bridges along the coast line will be elevated to reduce the likelihood of storm surge damage. In addition, vertical restraint devices and air vents may be used to reduce the chances of bridges having decks unseat and collapse.

1.1 ASCE-TCLEE Investigation Team

A reconnaissance team formed by TCLEE with support from the Mid-America Earthquake Center prepared this report. The period of the investigation was from November 6-10, 2005, with follow-up telephone and email communications.

This report was prepared under the direction of Curtis Edwards. The volunteer team consisted of the following people:

Reginald DesRoches (Associate Professor, Georgia Tech, Mid-America Earthquake (MAE) Center), Team Leader and Editor
Bryant Nielson (Assistant Professor, Clemson University, MAE Center)
Jamie Padgett (Graduate Research Assistant, MAE Center, Georgia Tech)
Nick Burdette (Graduate Research Assistant, University of Illinois, Urbana-Champaign, MAE Center)
Oh-Sung Kwon (Graduate Research Assistant, University of Illinois, Urbana-Champaign, MAE Center)
Ed Tavera (Geotechnical Engineer, Fugro West, Inc.)
Mark Yashinsky (Seismic Engineering, Caltrans)

1.2 Acknowledgements

The report would not have been possible without the assistant of many kind people in the gulf coast region, including the following:

- Steven Twedt, District 6 Construction Engineer, MDOT
- Paul Fossier, Bridge Design, LADOTD
- Gill Gautreau, Structures & Facility Maintenance Administrator, LADOTD
- Vince Latino, Fleet & Movable Bridge Maintenance Engineering, LADOTD
- Rhett Desselle, Change Management Facilitator, LADOTD
- Delvin Adamas, Mark Andrus, and Jason Faucheaux (LADOTD, District 02 Bridge Inspection)
- John Horn, Volkert and Associates (I-10 Twin Span Bridge Tour)
- Norfolk Southern Railroad
- Fred Conway, Chief Bridge Engineer of the Alabama Dept. of Transportation

2 Overview of Hurricane Katrina

Hurricane Katrina was the third most powerful storm of the 2005 Hurricane season, and the sixth-strongest Atlantic hurricane ever recorded. Katrina formed near the Bahamas on August 23, 2005, and crossed southern Florida (near Miami) as a category 1 storm before strengthening rapidly in the Gulf of Mexico, becoming the strongest hurricane ever recorded in the Gulf of Mexico (a record later broken by Hurricane Rita). In the Gulf, Katrina became a Category 5 hurricane with maximum winds of 175 mph and minimum central pressure of 902 mbar. The storm weakened considerably and made landfall as an extremely large Category 3 storm at 6:10 am CDT on August 29th along the Gulf Coast near Buras-Triumph, LA (approximately 40 miles southeast of New Orleans).

The physical size of Katrina caused devastation far from the eye of the hurricane. Heavy damage was inflicted onto the coasts of Louisiana, Mississippi, and Alabama, making

Katrina the most destructive and costliest natural disaster in the history of the United States and the deadliest since the 1928 Okeechobee Hurricane. The official death toll stood at 1417 (as of January 18, 2006), making it the fourth or fifth highest in US history. As of January, 2006, more than 3200 people remain missing.[1]

Over 1.2 million people in the Gulf Coast region were under an evacuation order before landfall, and by early September, 2005 more than 1.5 million people had been displaced. The overall economic damage from Katrina is estimated to be between $40-120 billion, almost double the previous most expensive Hurricane, Andrew, making Katrina the most expensive natural disaster in US history.

2.1 Wind Speed and Storm Surge in Gulf Coast

The damage to the infrastructure sustained during the storm was attributed primarily to the high wind speeds and storm surge along the coastal regions (aside from the flooding in New Orleans). Wind speeds exceeding 140 mph were recorded at landfall in southeastern Louisiana while winds gusted to over 100 mph in New Orleans, just west of the eye. As the hurricane approached the Mississippi/Louisiana border, wind speeds were approximately 125 mph. Gusts of over 80 mph were recorded in Mobile, Al. Hurricane Katrina produced extremely large storm surge heights in a region stretching from southern Louisiana to as far east as Mobile Bay, Alabama.

The initial storm surge estimates indicate that many parts of southern Louisiana, and the Mississippi gulf coast had storm surge heights much higher than would typically be considered in design. In Louisiana, the storm surge reached upwards of 25 ft near the location of the 5 mile I10 Twin spans. The storm surge resulted in the collapse of many of the spans of this bridge. In Mississippi, several areas reported storm surge heights of nearly 30 ft. The excessive storm surge was responsible for the collapse of two major bridges in the Mississippi gulf.

[1] http://www.nhc.noaa.gov/pdf/TCR-AL122005_Katrina.pdf

3 Overview of Damage

Approximately 45 bridges sustained damage in Alabama, Louisiana, and Mississippi. Most damaged bridges were adjacent to water, with the majority of the damage primarily occurring to the superstructure. Typical superstructure damage included unseating or drifting of decks and failure of guardrails due to storm surge. It was observed that the superstructure damage largely depended upon the connection type between decks and bents. For example, many bridges on I-10 either failed or had significant damage, while neighboring bridges on US-11 were not damaged. This discrepancy in bridge performance existed since the decks of US-11 were rigidly connected to piers while decks on I-10 were simply supported and did not have transverse shear keys. A similar trend was observed at US-90 (Biloxi-Ocean Springs) and a neighboring railroad bridge. The overall cost to repair or replace the bridges damaged during hurricane Katrina, including emergency repairs, and rebuilding in the Gulf Coast Region is estimated at over $1 billion.

Along the coastal area where storm surge was severe, many roads were heavily damaged and/or had significant deposits of debris, further hindering traffic and recovery efforts for several weeks. The cost for debris removal in the tri-state area was estimated at $200 million. Tables 3-1, 3-2, and 3-3 below summarize the damage to bridges during Katrina.

Table 3-1 Damaged bridges in Mississippi

Bridge	Carried	Bridge Type	Damage
I-10 Pascagoula River Bridge	I-10 at Moss Point	Prestressed Concrete Girder (Fixed)	Moderate damage to girders and span shifting due to barge and tug-boat impact. Repair costs estimated at $5.8 million.
US90 Bay St. Louis Bridge	US90 from Bay St. Louis to Pass Christian	Movable	Significant damage with bridge collapse due to unseating of many spans. Repair costs approximately $267 million.
US90 Henderson Point Bridges	US90	2 parallel precast, prestressed multi-span bridges (Fixed)	Significant damage with shifting of 6 spans, unseating of one, and moderate abutment damage. Repair costs approximately $1.9 million.
Biloxi-Ocean Springs Bridge	US90	Multi-span pre-stressed concrete girder with bascule (Movable)	Significant damage with bridge collapse due to unseating of many spans. Repair costs of $275 million.
Biloxi Back Bay Bridge	I-110	Pre-stressed concrete girder bridge with a movable bascule (Movable)	Moderate damage to piers due to barge impact. Minor damage to approach, abutments, and drawbridge arm. Repair costs of $2.5 million
David V. LaRosa Bridge	W. Wittman Road	Multi-span concrete girder (Fixed)	Moderate damage with to shifting of spans. Repair costs less than $60,000.
Popps Ferry Bridge	Popps Ferry Road	Movable	Significant damage with bascule system damaged, spans shifted, and bents, bearings, and girders damaged. Cost of repair approximately $7.7 million.

Table 3-2 Damaged bridges in Louisiana

Bridge Name	Carried	Bridge Type	Damage
Lake Pontchartrain - Orleans	I-10	Fixed	Significant - spans shifted/collapsed; bearing, girder, bent cap and pile damage. Emergency repair cost over $30 million.
Pontchartrain Causeway - Jefferson	LA-Causeway	Fixed	Significant - spans shifted/collapsed on turnarounds, abutment scour, approach slab undermining. Emergency repair cost over $1.5 million.
Caminada Bay - Jefferson	LA-1	Fixed	Significant - spans shifted, abutment scour, approach slab undermining. Emergency repair cost under $500,000.
Bayou Barataria - Jefferson	LA-302	Movable	Moderate – surge damage, submerged electrical/mechanical. Emergency repair cost over $50,000.
Harvey Canal - Jefferson	LA-18	Movable	Slight - damaged traffic gate, electrical. Emergency repair under $2,000.
Intracoastal Waterway @ Larose - Lafourche	LA-1	Movable	Moderate – wind damage to tower electrical cables and operator house. Emergency repair cost over $170,000.
Bayou Lafourche @ Leeville - Lafourche	LA-1	Movable	Significant – operational, structural and scour damage. Emergency repair cost over $1.6 million.
Lockport Company Canal - Lafourche	LA-1	Movable	Slight – wind damage to machinery housing. Emergency repair cost under $2,000.
Golden Meadow - Lafourche	LA-308	Movable	Moderate – wind damage to operator house roof. Barge stuck on fender system. Emergency repair cost under $9,000.
Galliano - Lafourche	LA-308	Movable	Slight – wind damage to operator house roof. Emergency repair cost under $5,000.
Inner Harbor Navigation Canal - Orleans	Florida Ave	Movable	Significant - barge impact to fenders, machinery submergence, approach slab undermining. Emergency repair cost over $500,000.
Seabrook - Orleans	Local Road	Movable	Moderate – wind damaged operator house. Fence and signal damage. Emergency repair cost under $25,000.
US 11 @ Lake Pontchartrain - Orleans	US-11	Movable	Significant – abutment erosion, bridge rail damage, damaged traffic gates and signals. Emergency repair cost over $6 million.
Chef Menteur - Orleans	US-90	Movable	Significant - eroded abutments, approach slab undermining, damaged electrical and mechanical, damaged fenders. Emergency repair cost under $2 million.
Rigolets Pass – Orleans	US-90	Movable	Significant - eroded abutments, approach slab undermining, damaged electrical and mechanical, damaged fenders. Emergency repair cost over $2 million.
Rigolets Pass - Under Construction - Orleans	US-90	Movable	Significant – Damaged girders and loss of construction material. Replacement cost over $1.7 million
North Draw - Lake Pontchartrain - Orleans	US-11	Movable	Moderate – pumping system submerged. Emergency repair cost over $50,000
Claiborne - Orleans	LA-39	Movable	Moderate – wind and water damage to operator house switch board . Emergency repair cost under $40,000.
Belle Chase - Plaquemines	LA-23	Movable	Moderate – Tower cable and shroud damage, window and handrail damage to operator house. Emergency repair cost over $200,000.

Doullut Canal - Plaquemines	LA-11	Movable	Significant – approach washout, submerged operator house, electrical and mechanical. Emergency repair cost under $700,000.
Perez – Plaquemines	LA-23	Movable	Moderate – Emergency repair cost over $200,000.
Yscloskey - St. Bernard	LA-46	Movable	Significant – approach washout, submerged operator house, electrical and mechanical. Emergency repair cost under $900,000.
St. Bernard Canal – St. Bernard	LA-46	Movable	Moderate – Emergency repair cost over $40,000
Bayou Des Allemands - St. Charles	LA-631	Movable	Slight - damaged traffic gates and navigation lights. Emergency repair under $3,000.
Tchefuncte River Madisonville - St. Tammany	LA-22	Movable	Moderate - submerged electrical and mechanical, abutment scour. Emergency repair cost over $25,000.
Bonfouca - St. Tammany	LA-433	Movable	Moderate - submerged electrical and mechanical. Emergency repair cost over $200,000.
Bayou Liberty - St. Tammany	LA-433	Movable	Significant – bent pivot arm, submerged electrical/mechanical. Emergency repair cost over $1.5 million.
East Pearl River - St. Tammany	US-90	Movable	Moderate - erroded approach slabs, submerged electrical/mechanical. Emergency repair cost under $400,000.
West Pearl River - St. Tammany	US-90	Movable	Moderate – roof blown off of operator house, damaged electrical and mechanical. Emergency repair cost under $350,000.
Presque Isle @ Bayou Petite Caillou - Terrebonne	LA-24	Movable	Slight - damaged operator house. Emergency repair cost under $1,000.
Country Club - Terrebonne	LA-3127	Movable	Slight - damaged operator house. Emergency repair cost under $1,000.
Bayou Dulac - Terrebonne	LA-57	Movable	Slight - damaged operator house. Emergency repair cost under $1,000.
Houma Navigation Canal - Terrebonne	LA-661	Movable	Slight - damaged operator house. Emergency repair cost under $1,000.

Table 3-3 Damaged bridges in Alabama

Bridge	Carried	Bridge Type	Damage
Cochrane Africatown USA Bridge	US 90	Cable-Stayed Bridge (Fixed)	Moderate damage to bearings, tower, and cable. Repair cost about $1 million.
Mobile Delta Causeway	I-10 to US90/98	Precast I-girder Connector (Fixed)	Major damage due to movement of simply-supported spans. Repair cost $1.14 million.
Dauphin Island Parkway Bridge	Highway 193	Precast Segmental Bridge (Fixed)	Major damage to fenders. Minor damage to approaches. Repair cost $6 million.
Bayou La Batre Bridge	Highway 188	Precast single-span bridge (Fixed)	Moderate damage to girders and approaches. Repair cost under $10,000.

4 EMERGENCY PREPAREDNESS

Emergency preparedness plans are critical for limiting the damage during hurricanes as well as facilitating rapid response. Below, a brief summary of the emergency preparedness plans for Mississippi, Alabama, and Louisiana is provided.

4.1 Emergency Preparedness in Mississippi

In 2001, the Mississippi Department of Transportation developed a "Comprehensive Emergency Response Plan" with a supplemental "Hurricane Response Plan." These plans were established because of the awareness of a fairly common hurricane threat to the Gulf Coast region, and in recognition of the potential risk to the transportation network. They outline responsibilities of different divisions within the DOT, as well cooperation with outside agencies, such as the National Weather Service, Mississippi Emergency Management Agency (MEMA), and Mississippi Department of Public Safety. The plans proved valuable in Hurricane Katrina for effective evacuation of the residents through hurricane evacuation routes (Figure 4.1), traffic control, and execution of the I-59 contraflow plan prior to the storm. A formal "Interstate 59 Contraflow Plan for Hurricane Evacuation" had been established in 2003, detailing the staffing and staging of personnel and coordination with the LADOTD.

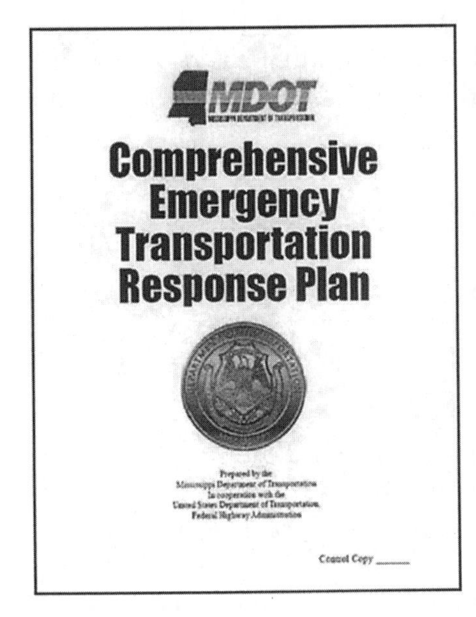

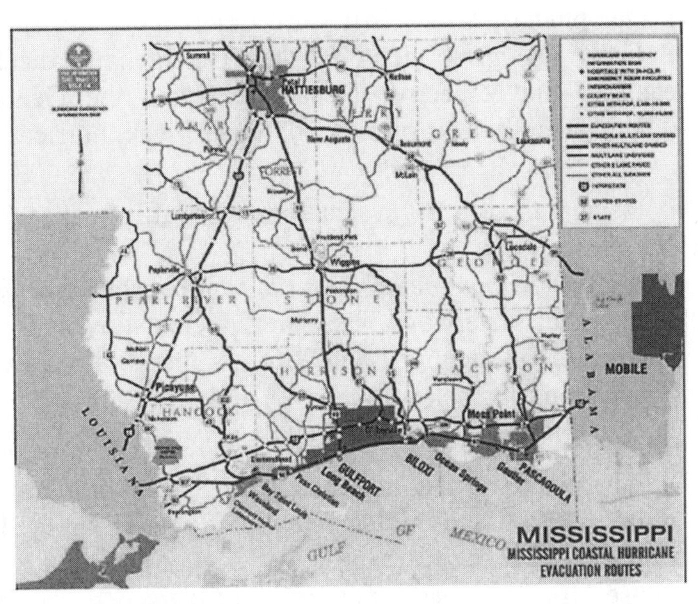

Fig. 4.1. MDOT Emergency Response Plan and Hurricane Evacuation Routes.

The emergency preparation also included stockpiling of additional equipment for emergency response in the Hattiesburg office, near the Mississippi Coast. This included equipment for immediate response such as evaluation of bridges and roadways and for clearing debris following the event. This stockpiling of equipment and plan for coordination with other offices and agencies proved effective in Katrina. Twenty-five inspectors from MDOT responded to the needs following Katrina, and were assisted by County foremen who began assessing damage the night after the storm, and later by Florida DOT bridge inspectors. There was a shortage of fuel experienced, which had to be acquired from private companies. Temporary housing for some workers who lost their homes in the storm was set up at the DOT headquarters, and most MDOT employees reported to work and were available to carry out the response efforts.

4.2 Emergency Preparedness in Louisiana

The Louisiana Office of Emergency Preparedness (OEP) manages federal and state disasters, both man-made and natural, including homeland security concerns. They provide communication and partnerships with the Military Department in New Orleans, the Governor's Office, the Legislature, Congressional staff, State officials, Parish and City officials, Parish Emergency Directors, individual citizens and FEMA.

The OEP in conjunction with other state agencies has prepared several documents to assist Louisiana citizens in emergency evacuations. The "Louisiana Citizen Awareness & Disaster Evacuation Guide" contains Louisiana Evacuation Maps and New Orleans Contraflow Evacuation Maps and instructions.

The *"Louisiana Emergency Evacuation Map"* was developed as a result of a working group of members of the LADOTD (Louisiana Department of Transportation and Development), OEP (Office of Emergency Preparedness), and LSP (Louisiana State Police). The group identified State Emergency Evacuation Routes that could be used by each of the respective Departments and the General Public for disaster situations (chemical spill, ice storm, flood, nuclear leak, hurricane, etc.).

Louisiana emergency evacuations for traffic management have been classified in three phases:
Phase I *–50 Hours before onset of tropical storm winds.* Includes areas south of the Intracoastal Waterway. These areas are outside any levee protection system and are vulnerable to Category 1 and 2 storms. These areas are depicted in RED on the Evacuation Map below.

Phase II- *40 Hours before onset of tropical storm winds.* Includes areas south of the Mississippi River that are levee protected but remain vulnerable to Category 2 or higher storms. These areas are depicted in ORANGE on the Evacuation Map. During Phase II, there are no route restrictions.
Phase III -*30 Hours before onset of tropical storm winds.* Includes areas on the East Bank of the Mississippi River in the New Orleans Metropolitan Area which are within the levee protection system but remain vulnerable to a slow-moving Category 3 or any Category 4 or 5 storm. These areas are depicted in YELLOW on the Evacuation Map.

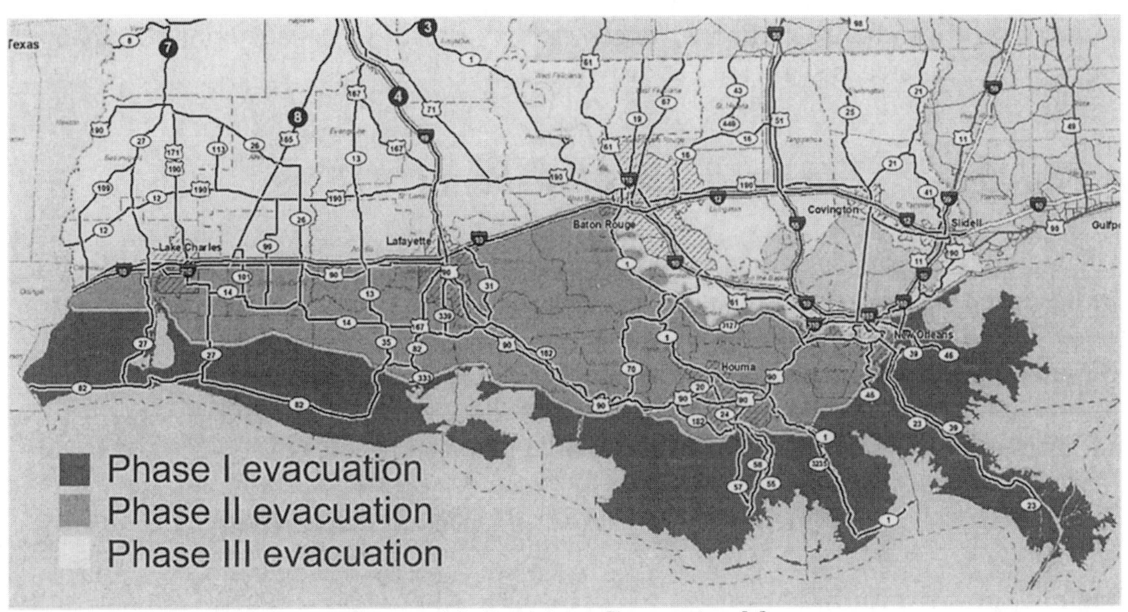

Figure 4.2. Louisiana Evacuation Map

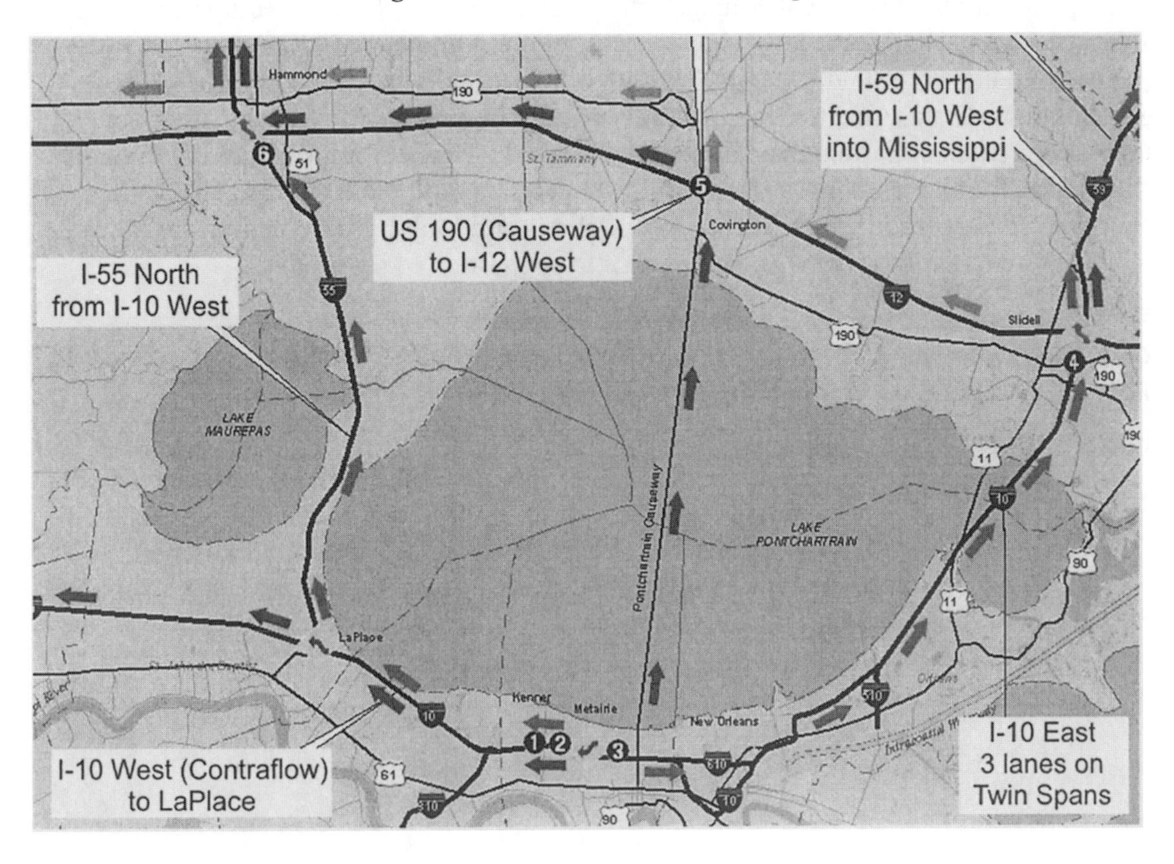

Fig 4.3. New Orleans Contraflow Map

Contraflow was put in place 3 – 5 years ago. This facilitated the closure of on ramps and reversing the direction of traffic flow on some roadways. The implementation of contraflow allowed for quick evacuation of the city for those who had the means to leave by using their own transportation. The contraflow map for management of traffic out of the New Orleans area is shown in Figure 4.3.

4.3 Emergency Preparedness of Alabama DOT

Although the Alabama DOT had no written emergency plan, they were well prepared for the hurricane because of the frequency and nature of large, destructive storms hitting the state. The Alabama DOT pre-positioned their highway workers and bridge inspectors just out of the area of inundation so they would be able to clean up the roads and re-open the bridges as soon as the storm passed. They also tried to make the workers as self-sufficient as possible by providing them with fuel, food and water.

4.4 Emergency Preparedness of Railroad Companies

Norfolk Southern Railway

In preparation for Hurricane Katrina, Norfolk Southern moved rolling equipment on its lines near coastal areas and low lying areas in southern Louisiana, Mississippi, and Alabama to higher ground. Operations south of Meridian, Miss., to New Orleans, La., and South of Selma, Ala., to Mobile, Ala., were discontinued early Sunday afternoon. Operations south of Birmingham also were curtailed. Traffic normally moving through these areas for interchange was rerouted in cooperation with other carriers. Norfolk Southern embargoed all shipments to New Orleans and Mobile. This action was taken to avoid further complication of operations due to congestion in the area.

CSX Transportation

Predictions were made that the storm would make landfall near New Orleans on mid-day Monday, Aug. 29. Floodgates in New Orleans were closed at 6 a.m. on Sunday, Aug. 28, and rail traffic in the area was suspended. CSXT monitored the path of Hurricane Katrina and activated its hurricane preparedness plan for the Gulf Coast. Some traffic scheduled for interchange with other railroads at New Orleans was rerouted through different gateways. As a result of the initial impact of Katrina in south Florida on Friday, Aug. 26, freight and passenger service on CSXT's network was restored incrementally as subdivisions were inspected and deemed safe for operations.

5 HIGHWAY BRIDGES PERFORMANCE AND REPAIR

5.1 Performance and Repair of Highway Bridges in Mississippi

As shown in Figure 5.1, the majority of the damage in Mississippi occurred either along the coast (US90), or along routes connecting I10 to US90. Most of the damage was confined to the region stretching from as far west as Bay St. Louis to as far east as Ocean Springs. In all, approximately 6 bridges had moderate-to-extensive damage. Most of the damage resulted from storm surge or impact from barges. Bridge closures ranged from only a few days to cases where the repair will take 2-3 years. Below is a detailed discussion of some of the cases of bridge damage in Mississippi.

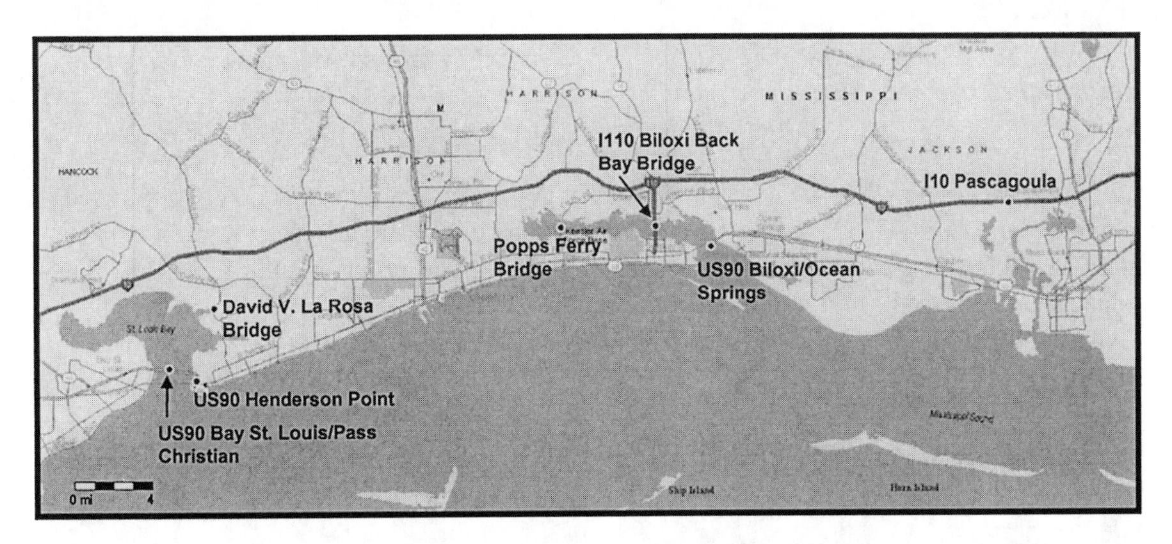

Fig. 5.1. Map Showing Damage to Bridges in Mississippi

US90 Henderson Point Bridges

Two parallel bridges carry US90 over the CSX railroad near Bay St. Louis in Harrison County, MS. These bridges are oriented in the north-south direction and carry two lanes of traffic each in the eastbound and westbound directions. The bridges are multi-span with pre-cast, pre-stressed bulb-T girders having elastomeric pads on the free end and doweled fixed ended diaphragm connections on the alternate supports. The storm surge in this area was reported to be on the order of 25-30', leading to uplift and shifting of a total of six spans on the Henderson Point bridges. Spans on the east-bound bridge had transverse movement in excess of 1.5' on the free end, and damage to the shifted diaphragm on the fixed end (Figures 5.2-5.3). The abutments suffered damage from pounding and transverse shifting of the superstructure, and there was significant washout of the backfill. The bridge carrying west-bound traffic had one span completely unseat. Figures 5.4-5.5 show the failed doweling used for a fixed connection between the span and bent. This bridge, along with the entire US90 route was inspected and closed on August 29, 2005.

Fig. 5.2 Transverse movement of span to eastbound section of US90 at Henderson Pt.

Fig. 5.3 Diaphragm damage on fixed end of US90 at Henderson Pt.

Repair of the US90 Henderson Point bridges (completed in February, 2006) used in-house plans from MDOT and consisted of replacement of the unseated span, repair of damaged bents and girders from superstructure shifting, and replacement of backfill. The replacement of backfill and mobilization of repair crews had been accomplished as of November 7, 2005. The estimated total repair cost for the two bridges is $1.94 million.

Fig. 5.4. Fixed connection at unseated span of west-bound bridge of US90 at Henderson Pt.

Fig. 5.5. Span with repairs initiated at west-bound bridge of US90 at Henderson Pt.

Biloxi-Ocean Springs Bridge (US90)

The 1.6 mile Biloxi-Ocean Springs Bridge which carries four lanes of US90 between the two cities over the Biloxi Bay suffered complete damage during Hurricane Katrina. While this bridge had a movable section, the damaged spans included the lower elevation multi-span pre-stressed concrete girder sections near the ends of the bridge. Numerous spans were shifted and unseated (Figures 5.6-5.7) due to the storm surge which rose to levels in excess of 20' at the bridge site. In addition to the unseating of the spans, damage to the steel bearings was evident, as well as damage to the abutments with washout of backfill and settlement of the approach slabs. The bearings were steel sliding bearings with bronze cores which provided no apparent positive connection between the substructure and superstructure, as seen in Figure 5.8.

The extensive damage to the Biloxi-Ocean Springs Bridge led to immediate closure of the bridge and a need for complete replacement. The project is to be let as a design-build contract with a goal of partial completion by May, 2007, when one lane of traffic should be open in each direction. Full completion is targeted at November, 2007. The new bridge is to be a high-rise bridge (approximately 85' high) to avoid storm surge issues and eliminating the need for a movable span. The new bridge will have a capacity of 6 lanes and an added bike path. The Biloxi-Ocean Springs replacement is estimated at a total cost of approximately $275 million.

Fig. 5.6 East end of Biloxi-Ocean Springs Bridge

Fig. 5.7. West end of Biloxi-Ocean Springs Bridge

Fig.5.8 Damage to steel-bronze bearings due to transverse motion of bridge.

Biloxi Back Bay Bridge

The Biloxi Back Bay Bridge carries two north-bound and two south-bound lanes of I-110 over the Biloxi Back Bay. Following Hurricane Katrina, I-110 served as the only primary route to Biloxi, due to the closure and damage to US-90 and the Biloxi-Ocean Springs Bridge. The Biloxi Back Bay Bridge is a pre-stressed concrete girder bridge with a movable bascule, as seen in Figure 5.9. The bridge was damaged due to barge impact, shearing the easternmost pile of the bent directly south of the bascule, which supports a span of pre-stressed girders as well as the bascule anchor span. Other minor damage was sustained, such as guardrail, sidewalk, and drawbridge gate arm damage, as well as differential settlement of the north approach and abutment.

The bridge was inspected on September 1, 2005 and the outside north-bound lane (over the damaged pile) was closed. The other three lanes remained open to traffic and allowed I-110 to serve as a critical route for the recovery efforts in Biloxi. Repair of the damage to the Biloxi Back Bay Bridge was performed using in-house repair plans and completed within 38 days following Katrina. The cost of the repairs was $2.5 million.

Fig. 5.9. Pile section impacted by barge of Biloxi Back Bay Bridge

Fig. 5.10. Close-up of damaged pile of Biloxi Back Bay Bridge

I-10 Pascagoula River Bridge

Two twin structure, 300-plus span bridges carry I-10 over the Pascagoula River near Moss Point, MS. The east-bound bridge was impacted by a construction barge during Hurricane Katrina, as seen in Figures 5.11-5.12. The storm caused surges ranging from 13'-18' in the area. The impact caused a 3.75' misalignment of a 6-span (312') continuous unit of the bridge. The piles of the five bents supporting this span were sheared and one span of fascia girder of the unit was completely destroyed. Impact from the tug-boats led to damage of other spans, where several fascia girders were destroyed or partially damaged due to spalling. Additionally, exposure of some pre-stressing strands and broken strands were visible in some spans.

Fig. 5.11. Barge impact on I-10 Pascagoula River Bridge (Courtesy of MDOT)

Fig. 5.12 Forty-five inch transverse shifting of 6-span unit on I-10 Pascagoula River Bridge (Courtesy of MDOT)

Fig. 5.13. Bearing damage on I-10 Pascagoula River Bridge due to barge impact.

Fig. 5.14. Pile damage on I-10 Pascagoula River Bridge due to barge Impact.

The bridge was closed on August 29, and inspected thoroughly the following day. Crossovers were built to move the east-bound traffic over to the twin structure such that it could facilitate one lane of both east- and west-bound traffic. This cross-over was constructed by local contractors and performed within 8 days of the bridge closure. Barge removal from the marsh was necessary prior to the repair, which required excavation and pumping to allow the barge to float and navigate away from the site. The existing plans were used for the repairs (Figure 5.15). MDOT had pre-cast piling and beams fabricated immediately following the inspection and furnished them to the contractor. The repair of the bridge was contracted with a bonus incentive of $100,000/day of completion prior to 30 days. The repairs were completed in approximately 20 days by October 2, 2005, as shown in Figure 5.16, with a total cost of $5.8 million.

Fig. 5.15 Repair of I-10 Pascagoula River bridge

Fig. 5.16. Replaced 6-Span Unit and Repaired bent cap and Piling of I-10 Pascagoula River Bridge

Bay St. Louis Bridge (US90)

The Bay St. Louis Bridge connects Hancock and Harrison counties across Bay St. Louis on US90 in Mississippi. Decks of the bridge have four lanes of roads. Each span consists of two decks and each deck has two lanes. Decks are simply supported by high-type steel bearings and each bent consists of seven piers and one cross beam.

After Hurricane Katrina, almost all decks of the bridge unseated. In the west part of the bridge, all decks completely fell off and submerged into water. Many spans of the North decks (running towards the west) at the East part of the bridge were had fallen off. South decks (running towards the east) at the East part of the bridge drifted to the North and partially submerged. Many bearing-pedestal connections and bearing-deck connections were damaged due to high transverse load, Figures 5.17-5.18. The bridge was closed after the hurricane.

The bridge is planned to be completely replaced with a new bridge, which will carry four lanes and have a peak vertical clearance of 85' for marine traffic. Approach spans will have 35' of clearance which is considered to be above storm surge levels. Decks will be built as cast-in-place reinforced concrete T-beam. Estimated cost is approximately $267 million. The project is planned to be completed by June 30, 2007.

Fig. 5.17. Close-up of East End of US90 Bay St. Louis Bridge

Fig. 5.18. Bearing Damage on US90 Bay St. Louis Bridge

Popps Ferry Bridge

The Popps Ferry Bridge is a bascule bridge located in Biloxi, Mississippi. The two-lane roadway is a key artery for accessing the densely populated housing areas north of the Biloxi peninsula and carries over 20,000 vehicles a day when in operation [2].

During the storm, the mechanical house went under water, damaging the electrical system and navigation controls for the bascule. Due to the damage of the movable bridge, many boats were stuck within Big Lake. Several spans of the bridge were shifted during the storm, due to dowel failure. In addition, a few of the bents were damaged. After the hurricane, the bridge was closed which severely affected local traffic. People needed to detour 37 miles out of the way to get to the other side of the bridge.

The repair project was completed within 100 days with an incentive of $50,000 for every day when the project finish earlier than contracted and $50,000 of penalty per day of delay. The repair project included complete replacement of 8 caps, 12 spans, and 1 partial span. Seven spans were jacked back into place that were shifted up to 5' due to storm surge. Epoxy was grouted into cracks in caps that were being saved.

Fig. 5.19. Drifted Spans on Popps Ferry Bridge

Fig. 5.20. Repair of Bridge Damage on Popps Ferry Bridge

[2] Referred from http://www.gulfcoastnews.com/GCNnewsPoppsFerryRepairAhead.htm.

David V. LaRosa Bridge

The David V. LaRosa Bridge carried traffic along W. Wittman Road in Harrisson county, Mississippi, and is owned by the local district. During the storm, a few spans had drifted. The drifted spans were jacked back into their original location and the bridge became fully functional 60 days after hurricane.

Fig. 5.21 David V. LaRosa Bridge

5.2 Performance and Repair of Highway Bridges in Louisiana

Highway bridges in Southeastern Louisiana were subjected to high winds and severe storm surge levels during hurricane Katrina. While the majority of the bridges in the affected area received little or no damage, there were approximately 25 bridges that sustained moderate-to-significant levels of damage. One report by the Louisiana Department of Transportation and Development (LADOTD) stated that 22 of the 86 movable bridges in the affected area were noticeably damaged. The other three damaged bridges are of a fixed type situated over open bodies of water. Figure 5.22 geographically locates some of the damaged bridges in the region.

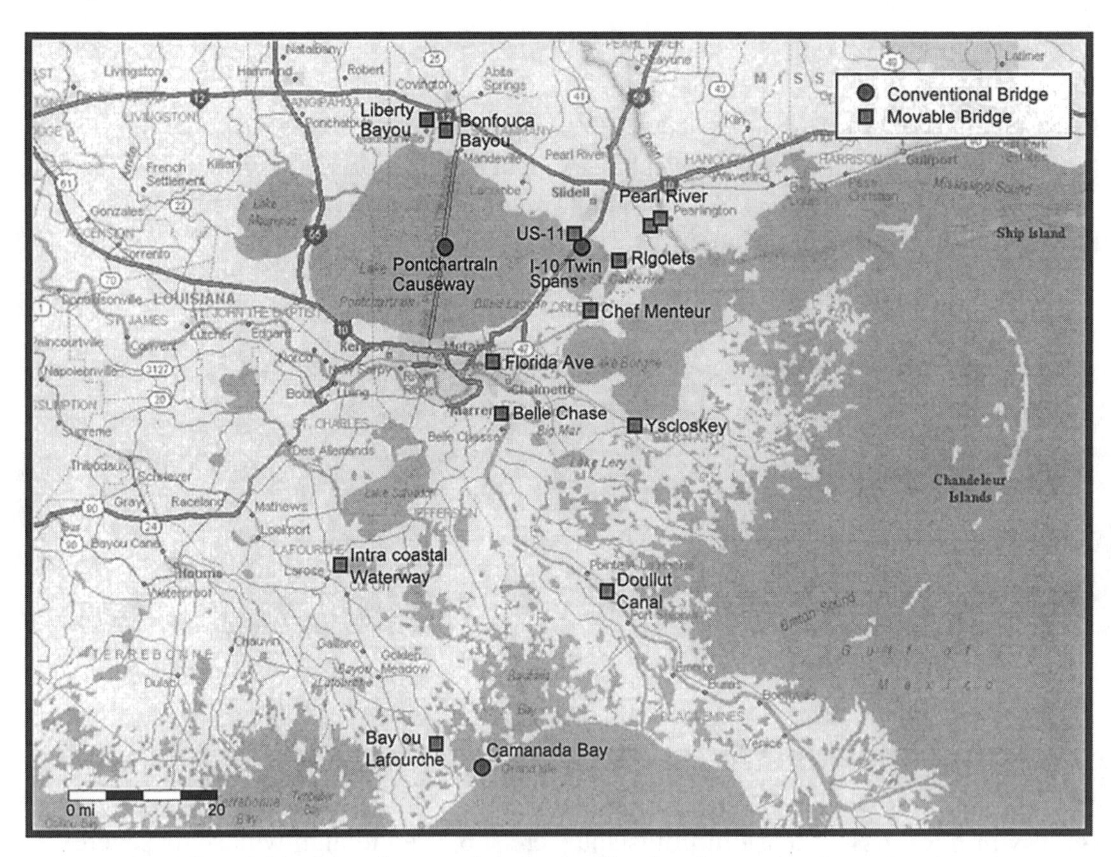

Fig. 5.22. Map showing locations of bridge damage in Louisiana.

I-10 Twin-Spans

The greatest extent of damage to any one bridge in Louisiana occurred to the I-10 twin-spans over Lake Pontchartrain. This bridge consists of two separate 5.4 mile long spans, each providing two lanes of traffic between St. Tammany and Orleans parishes. Built in 1963, this bridge was constructed using design details typical of bridges in the region. The main bridge consists of three steel girder spans positioned on multi-pile bents elevated high above the water level to accommodate marine traffic (Figures 5.23-5.24). The 433 simply supported approach spans are constructed of pre-cast prestressed concrete segments. Each 65 foot segment has six girders cast monolithically with the concrete deck. The concrete spans bear on three-pile bents using steel and bronze bearings, placing the top of the bridge deck at approximately 15 feet above the mean water level. The piles are 54 inch diameter and approximately 150 foot long prestressed hollow cylinders and are capped with a reinforced concrete beam.

Fig. 5.23. Steel girder spans of main bridge in I-10 Twin Spans.

Fig. 5.24. Pre-cast concrete approach spans in I-10 Twin Spans.

Damage to the I-10 twin spans was largely due to storm surge and occurred primarily in the low lying approach spans, leaving the elevated sections of the bridge in good condition. The damage, which was in the form of individual spans shifting off their bearings, occurred through a combination of buoyant forces and lateral water hammering. The LADOTD calculated that each span weighs approximately 500 kips yet the surge produced uplift forces around 900 kips. Once the individual spans were lifted off their bearings, wave and wind forces caused many spans to fall in the water. Figures 5.25-5.26 show schematics provided by the LADOTD illustrating the sequence of events. This resulted in a total of 473 spans which were shifted and an additional 64 spans in the water (See Table 5-1). Figure 5.27 depicts the nature of the damage showing both shifted and displaced spans as well as missing barrier railing. The displaced spans generally failed at the bearings due to excessive corrosion in the shear studs. This generally served to preserve the integrity of the bearing base plates and anchor bolts yet often resulted in significant damage to the girders as shown in Figures 5.28-5.29.

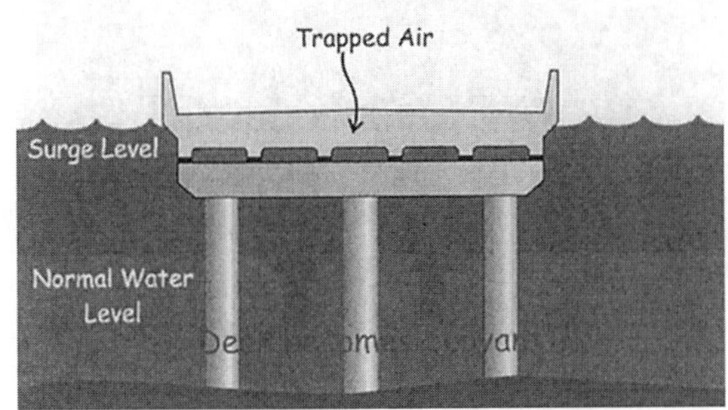

Fig. 5.25. Schematic of mechanism causing span displacement - buoyant forces

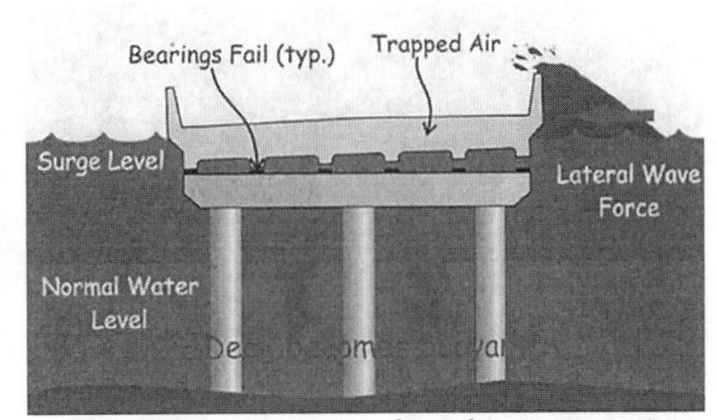

Fig. 5.26. Schematic of mechanism causing span displacement - lateral forces

Table 5-1 Summary of Damage to I-10 Twin-Spans

Element	Westbound	Eastbound
Spans in water	26	38
Spans shifted	303	170
Feet of barrier rail missing	13910	130
Bents missing	1	0

Fig. 5.27. Typical Damage to I-10 Twin-Spans (Courtesy of LADOTD).

Fig. 5.28. Bent beam and bearing base plates (Courtesy of LADOTD)

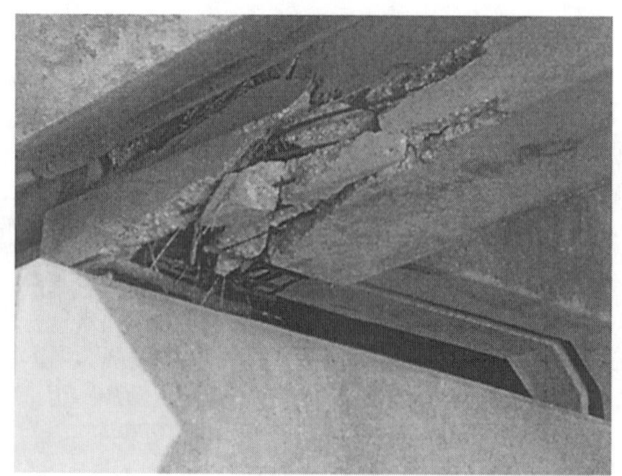

Fig. 5.29. Fractured girder seat (Courtesy of LADOTD)

The repair and recovery of the I-10 Twin-Spans was fast tracked due to its importance as a major transportation link. A thorough survey and documentation of the damage was conducted on September 2-4. In addition, representatives from the Florida DOT were consulted on bidding and repair procedures because of their recent experience in 2004 with the I-10 bridge over Escambia bay near Pensacola, Florida. These measures allowed the letting of the repair bid to occur 7-8 days after Hurricane Katrina. A local company was awarded the contract for the repair work with a bid of $30.9 million. The contract was to be carried out in three phases. The first phase was for the reopening of the eastbound bridge span by October, 2005, with an incentive/disincentive program of $75,000/day. The second phase was to reopen the westbound span with an incentive of $75,000/day and a disincentive of $25,000/day. The third phase is to provide daily inspection and maintenance for a period of three years after completion of the repairs. The bid for a replacement bridge is scheduled to be let in the spring of 2006 with an estimated cost of $600 million, with completion within 3-4 years. Therefore current repairs must be made for an expected life of three to four additional years.

Repairs to the eastbound span were made by taking undamaged deck segments from the westbound span to replace extensively damaged or missing segments. The remaining concrete segments were simply realigned using barges and self propelled motorized transports (SPMT) as seen in Figures 5.30-5.31. Bronze girder bearings were either repaired with Teflon pads or completely replaced with elastomeric pads. The girders of many of the deck segments sustained minor to moderate damage. To minimize the number of segments which needed to be replaced, various repair strategies were implemented to accommodate the different degrees of damage. For example, when only minor damage was present at the girder seats, jacks were added under the diaphragms to assist in transferring loads to the bent (See Figure 5.32). When the damage to the girder seat was more extensive, bent saddles (Figure 5.33) or helper bents (Figure 5.34) were used. Using these strategies, the eastbound span was reopened on October 10th, which was day 28 of the contract. This was 15 days ahead of schedule resulting in a bonus of $1.1 million for the contractor.

Fig. 30. Span being swapped from westbound to eastbound (Courtesy of LaDOTD)

Fig. 31. SPMT for realigning deck segments (Courtesy of LaDOTD)

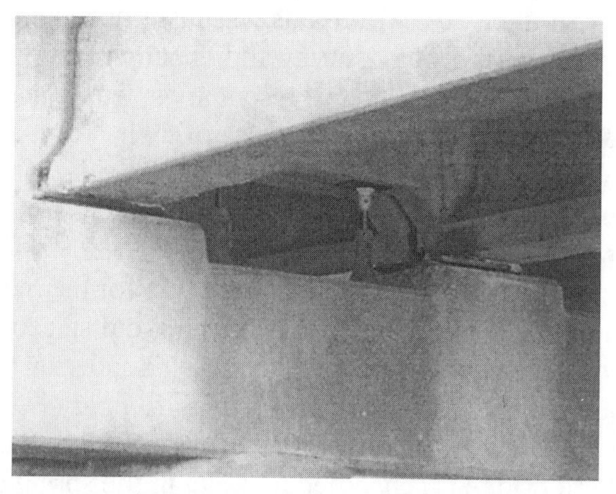

Fig. 5.32. Screw Jacks Placed Under Diaphragms.

Fig. 5.33. I-10 Repairs - Bent saddle

Fig. 5.34. I-10 Repairs - Helper bent

By the time phase I was completed, phase II was already underway. The repair procedures used for the westbound span were the same as those used for the eastbound span. However, several additional issues needed to be addressed. First, there was almost 14,000 feet of barrier railing which was damaged. This barrier railing had to be removed (see Figure 5.35) and replaced with concrete Jersey barriers. The second issue was that many deck segments needed to be replaced. Timing and resources did not allow for the manufacturing of new precast concrete segments. Therefore, modular steel bridge panels were used for the replacements (Figures 5.36-5.37). Approximately 4000 feet of steel bridge panels were needed to complete the project. Work on phase II proceeded rapidly and was completed nine days ahead of schedule on January 5, 2006.

Fig. 5.35. Removal of Damaged Barrier Rail.

Fig. 5.36. Modular steel bridge panels for I10 repair

Fig. 5.37. Steel bearings used to support modular steel bridge panels for I10 repair

Pontchartrain Causeway

The Pontchartrain Causeway is a 24 mile long twin-span bridge owned and operated by the Greater New Orleans Expressway Commission. It was built in 1956 to cross over the center of the lake and normally has an average daily traffic volume over 30,000 cars per day. Construction type for the causeway is very similar to that for the I-10 twin spans, using simply supported precast deck segments placed on multi-pile bents.

The damage to the causeway was not as extensive as for the I-10 twin-spans. The abutments for the southbound roadway were eroded as shown in Figure 5.38. The majority of the bridge segments were undamaged because the storm surge was not as high at this location as it was at the I-10 twin-spans. However, 17 spans were lost from the turnarounds, as shown in Figure 5.39, because of their elevation with respect to the mean water level.

Repair work for the causeway occurred in two phases. The first or emergency phase was to repair the abutments and provide erosion control measures. These repairs are temporary and are intended to get traffic flowing again. The estimated cost of these repairs is approximately $111,000. Phase II is to conduct permanent repairs including the replacement and realignment of spans in the turnaround as well as providing more permanent repairs to the approach slabs and abutments. This phase carries an estimated price tag of $1.4 million.

Fig. 5.38 Eroded approach slab on Pontchartrain Causeway (Courtesy of LADOTD)

Fig. 5.39 Missing spans in turnaround on Pontchartrain Causeway (Courtesy of LADOTD)

US-11 over Lake Pontchartrain

The US-11 bridge over Lake Ponthartrain is a 4.7 mile long single-span bridge over Lake Pontchartrain built in 1938. It is located in the proximity of the I-10 twin-spans but is of slightly different construction. The majority of the deck segments are haunched concrete girders which are continuous over multi-pile bents as shown in Figure 5.40. An interesting feature of the construction was the presence of air vents in the diaphragms located over the bents (see Figure 5.41).

Fig. 5.40. US-11 over Lake Pontchartrain general construction

Fig. 5.41. Diaphragm air vents on US-11 over Lake Pontchartrain

The damage to this bridge was relatively minor as compared with the I-10 twin-spans. The continuous nature of the spans is believed to be a major factor in the bridge sustaining little damage compared with the I-10 twin span. The vent holes, which help mitigate buoyant forces, are also believed to be a significant reason for the relatively small amount of damage. Aside from damage at the draw bridge, other damage was limited to erosion at the approach ways. The estimated cost for emergency repairs, which were completed around September 20, 2005, is approximately $327,000.

LA-1 Over Camanada Bay

The bridge over Camanada Bay was built in 1961 and carries state road LA-1 over to Grand Isle. The bridge is approximately seven tenths of a mile long and is constructed of simply supported reinforced concrete bridge segments. The storm surge caused 13 of the 110 simple spans to shift. Additionally, a moderate amount of damage was also sustained at the approach slab. Because of the generally poor condition of the bridge prior to Hurricane Katrina, it was more susceptible to localized damage from impacting debris. This damage was seen in the form of spalled concrete and exposed rebar. Emergency repairs realigning the deck segments and repairing the approach ways were completed by October 15, 2005 at a cost of approximately $437,000. Traffic flow was restored but load restrictions were recommended until further evaluation could be completed. The bridge is currently scheduled for replacement under the Federal Bridge Replacement Program.

Figure 5.42. Misaligned Spans of Grand Isle Bridge at Camanada Bay.

5.3 Performance and Repair of Highway Bridges in Alabama

Bridge damage in Alabama was much less severe than in the states of Mississippi and Louisiana. Hurricane Katrina caused moderate to major damage to four bridges in Alabama (Figure 5.43). However, Alabama has not always been so fortunate during hurricanes. In 2004, Hurricane Ivan caused significant bridge damage and in 1979, Hurricane Frederick slammed into Mobile Bay with 125 mph winds and washed away the original Dauphin Island Bridge, requiring the Alabama DOT to provide ferry service until a new bridge was built. In fact, the Alabama DOT was still providing ferry service between Dauphin Island and Fort Morgan until it was damaged during Hurricane Katrina

(Figure 5.44). The ferry had just returned to service after being damaged by Hurricane Ivan, which resulted in the state taking over ferry operations. With the loss of the ferry, travelers along the coast are forced to drive around Mobile Bay.

This section provides a description of the four Alabama bridges (along with their damage and repairs) that were impacted by Hurricane Katrina.

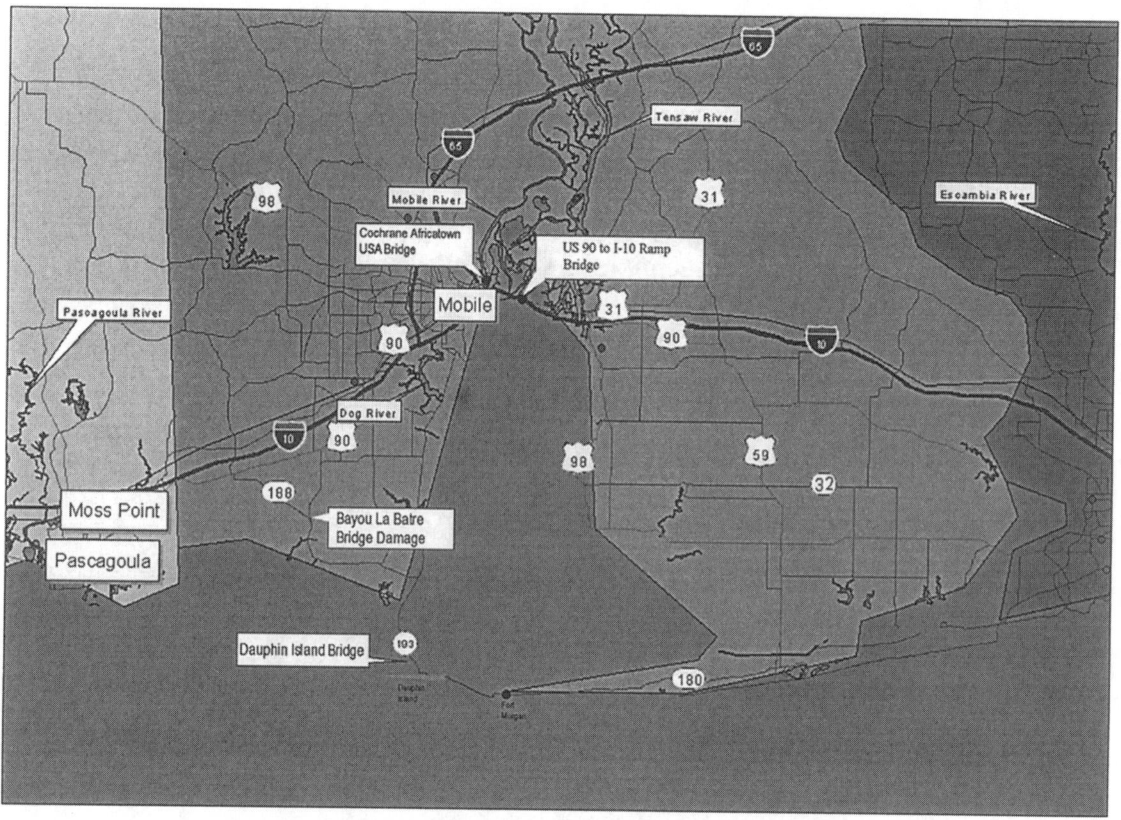

Figure 5.43. Map showing locations of bridge damage in Alabama.

Cochrane-Africatown USA Bridge

The Cochrane-Africatown USA Bridge (Figure 5.45) was built in 1991 and carries U.S. Highway 90 traffic over a port, railroad tracks, and the Mobile River. This reinforced concrete box girder bridge rises on rectangular, two- column bents to a height of 140 ft. and spans the river with a 780 ft long, cable-stayed main span. The bridge supports two lanes of traffic in each direction. The main span is supported by two, reinforced concrete, two-legged towers. The cable stays are arranged in a harp pattern along the outside edges of the deck and are connected to Taylor damping devices to reduce vibrations.

This bridge should have been high enough to have avoided debris thrown at it by the hurricane, but a 13,000-ton oil drilling platform, the PSS Chemul, used to house workers in the Gulf of Mexico, was being refitted at a local shipbuilding & repair facility on Blakeley Island when hurricane force winds pushed it from its mooring northward until it struck the bridge (Figure 5.46).

Considering the amount of damage that occurred, it is remarkable that the bridge remained in service following the hurricane. The platform smashed into the deck overhang, completely crushing the concrete barrier rail and striking one of the cable stays, but there was no apparent damage to the strands (see Figure 5.47). Large pot bearings support the superstructure at the towers and many of these were damaged by the collision (Figure 5.48). The concrete tower legs were also cracked. However, this bridge continued to carry traffic (two lanes with weight restrictions) after the storm. A $1 million contract was written by the Alabama DOT to repair the damage while allowing traffic to continue across the bridge.

Fig. 5.44. Fort Morgan Ferry in Mobile getting repaired following Hurricane Katrina (Courtesy of ADOT.)

Fig. 5.45. The Cochrane Africantown USA bridge in Mobile, AL.(Courtesy of MCEER/O'Connor).

Fig. 5.46. Oil Drilling Platform Blocked Impacting Cochrane Africatown Bridge (Courtesy of ADOT.

Fig. 5.47. Damage to cables on Cochrane Africatown bridge from impact with oil drilling platform (Courtesy of ADOT).

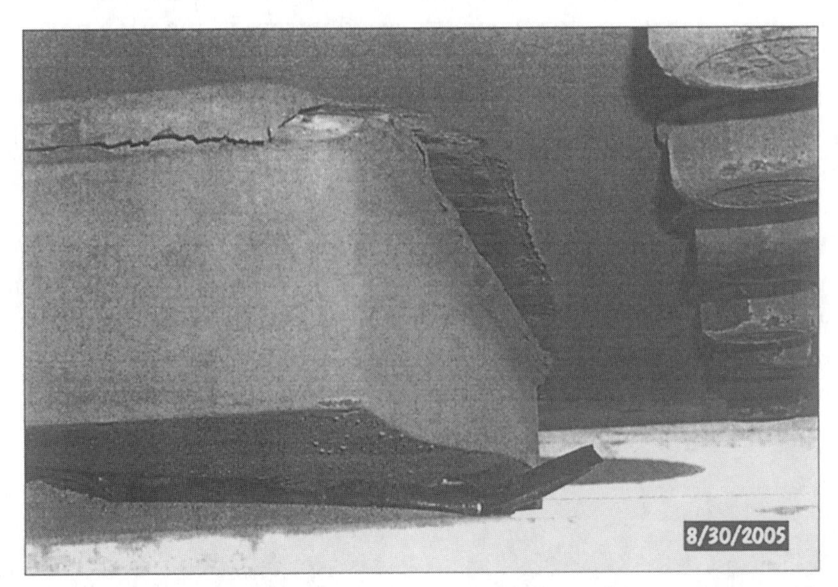

Fig 5.48. Bearing damage from impact with oil platform (photo courtesy of ADOT).

I-10 On-Ramp at Mid-Bay Crossing of US-90/98

This long (30 span), curved, single lane on-ramp was built in the early 1970's. It carries eastbound US 90 traffic over the north end of Mobile Bay and has 45 foot long, simply-supported, precast I girder spans (Figure 5.49) on sloping, rectangular column bents with dropped bent caps and driven pile foundations.

The storm pushed against the south side of the bridge, breaking the 1" diameter anchor bolts and the 6" x 9" x 1" steel angles that provided transverse restraint to the girders (Figure 5.50). This allowed the girders to move to the north and resulted in damage to five spans as well as the bent caps that supported them. Because the bridge is on a curve, the spans were shoved together as they moved laterally, which may have prevented them from being pushed off the bent caps. Recorded wind speeds in Mobile Bay were about 80 mph. It was the higher spans to the east that moved, while lower spans to the west remained undamaged. That may help explain why the spans were not washed off the bents during the storm.
This bridge closed in January of 2006 for six months while the damage is repaired. During that time drivers will have to use the on-ramps east and west of this location.

Fig. 5.49. Shifted at mid-bay crossing of US-90/98(Courtesy of MCEER/O'Connor)

Fig. 5.50. Damage to onramp of US-90/98 at bent (Courtesy of ADOT)

Precast bridge damage in Bayou La Batre

Most of the remaining bridge damage in Alabama was to the approaches. However, a one-span precast girder bridge had some spalling between the precast segments, apparently because there was no cast-in-place deck or diaphragm to keep the girders from banging together. Also, the approach embankments were too damaged to allow driving onto the bridge (Figures 5.51-5.52). However, this damage was repaired by road crews a few days after the storm.

Fig. 5.51. Approach to Bayou La Batre Bridge (photo courtesy of ADOT).

Fig. 5.52. Deck of Bayou La Batre Bridge (photo courtesy of ADOT).

Dauphin Island Bridge

As was previously mentioned, Hurricane Frederick destroyed the Dauphin Island Bridge in 1979. The new bridge was quickly designed and (using precast segmental construction) was re-opened to traffic in July of 1982. The main span is 400 feet, which made it the longest precast bridge span at the time. The bridge is 17,814 feet long and supported on tall, hollow piers and driven piles. Although the bridge was hammered by strong winds and a 20 ft storm surge, the structure survived Hurricane Katrina without much damage (Figure 5.53). The only damage was to the approaches (similar to the

Bayou La Batre Bridge) and to the fenders that protect the piers, which had to be rebuilt after the storm at a cost of $6 million (Figure 5.54).

Fig. 5.53. Photo of Dauphin Island Bridge after Hurricane (Courtesy of ADOT)

Fig. 5.54. Damage to approach span of Dauphin Island Bridge (Courtesy of ADOT)

5.4 Performance and Repair of Movable Bridges in the Gulf Coast

To support the significant marine and vehicular traffic in Louisiana, the state meets its transportation needs using 152 movable bridges—more than any other state. In a press release in mid-October, LADOTD Secretary, Johnny Bradberry, summarized storm damage noting that 22 of the 86 moveable bridges in the area affected by Hurricane Katrina had sustained significant damage. 17 of these structures are owned and operated by the state. The five other bridges are owned by local authorities, which in many cases are aided by the state as they request federal repair funds. Table 5-2 shows repair cost estimates, both emergency and permanent, for 16 of the state-owned damaged bridges, based on available LADOTD Damage Inspection Reports. The early estimates of damage to movable bridges showed that about $10 million would be needed for emergency repairs, which is consistent with summary Table 5-2. These emergency repairs will make each bridge functional, but further permanent repairs are needed to ensure safe long-term operation.

Table 5-2 Summary of repair costs for state-owned movable bridges damaged in Louisiana by Hurricane Katrina.

Route Number and Bridge Name		Movable Bridge Type	Debris Removal and Emergency Repair Cost*	Permanent Repair Cost*	Total Cost
US 90	Rigolets Pass	Swing	$3,046,450	$1,815,000	$4,861,450
US 90	Chef Menteur	Swing	$1,918,375	$1,210,000	$3,128,375
LA 433	Bayou Liberty	Swing	$1,510,000	$0	$1,510,000
US 90	East Pearl River	Swing	$409,200	$71,000	$480,200
LA 433	Bonfouca	Swing	$214,500	$0	$214,500
LA 302	Bayou Barataria	Swing	$57,000	$0	$57,000
LA 22	Tchefuncte River, Madisonville	Swing	$27,600	$0	$27,600
LA 46	LaLoutre (Yscloskey)	Lift	$895,500	$15,000	$910,500
US 90	West Pearl River	Lift	$350,000	$10,000	$360,000
LA 23	Perez	Lift	$221,000	$14,200	$235,200
LA 1	Larose	Lift	$170,960	$0	$170,960
LA 39	Claiborne	Lift	$39,300	$71,000	$110,300
LA 308	Golden Meadow	Lift	$13,600	$150	$13,750
LA 308	Galliano	Lift	$5,200	$900	$6,100
US 11	North Draw	Bascule	$358,971	$0	$358,971
LA 46	St. Bernard Canal	Bascule	$42,300	$0	$42,300
		TOTALS:	$9,279,956	$3,207,250	$12,487,206

* Repair cost estimates based on LADOTD Damage Inspection Reports

Because movable bridges affect not only vehicular traffic but important marine traffic as well, damage to moveable bridges in Louisiana has complicated travel routes for both modes of transportation. In many cases, water inundation destroyed lift motors and electrical systems, rendering structurally sound bridges immovable. Due to the importance of shipping to the region and marine transport of relief goods for disaster recovery, bridges with damaged mechanical components were often forced open to allow marine traffic to pass. The opening of the Chef Menteur Bridge at the entrance to Lake Pontchartrain is an example of this practice, with continued marine access to the lake breaking a vehicular link between the New Orleans area and St. Tammany Parish.

While much of the damage to movable bridges in Louisiana was similar regardless of movable bridge type, some was type specific. There are three main types of movable bridges and all three types were damaged in Hurricane Katrina. Moveable bridges can be: (i) *swing bridges* which have a center pivot pier about which a section of the superstructure rotates, opening two navigation channels to each side of the pivot pier, (ii) *lift bridges* where a central deck section is lifted vertically by two towers to open a navigation channel, and (iii) *bascule bridges*, which have one or two sections of superstructure that rotate up like a hinge, providing an opening for marine traffic. To provide representative examples of movable bridge damage and repair, one of each bridge type will be discussed in greater detail hereafter.

Swing Bridge: Chef Menteur Bridge

Though the Chef Menteur Bridge (Figure 5.55) is scheduled for replacement, this process could take as long as 10 years, and short-term repairs to the existing structure were needed immediately. An estimated $1.9 million was needed for debris removal and emergency repairs, with an additional $1.2 million for permanent fender system replacement. The emergency repairs included $500,000 for electrical and mechanical work on the movable span. Most of this money went towards replacing the electrical control system, but replacement of navigation lights, the operator house roof, and traffic gate arms was also needed. The most significant repair will involve removal of the five north approach spans affected by the slope failure. Because the foundation of their supporting bents has been compromised, the bents were removed and the multi-span gap bridged by a 126 foot long girder-supported deck. The current estimate for this repair was estimated at $1 million.

Fig. 5.55. Chef Menteur Bridge viewed from the south approach. Parapet damage from a boat collision during Hurricane Katrina can be seen in the photo.

Lift Bridge: Yscloskey Bridge

The Yscloskey Bridge on Route LA 46 in St. Bernard Parish was the lift-type bridge which sustained the most damaged from Hurricane Katrina. The elevated water raised the movable bridge deck 8', causing it to become skewed and stuck as the water level returned to normal (Figures 5.56-5.57). The high water level also submerged the electronics and bridge control system in the operator house for an extended period, ultimately destroying it. In an interview with Ed Douglas, an LADOTD Bridge Design Electrical Engineer, he explained that electrical damage was very common for movable bridges in the hurricane path. While most systems are designed for temporary wetting or submersion, extended submersion and rushing flood waters often destroyed bridge electronics.

Fig. 5.56. Yscloskey movable lift bridge on LA 46 - skewed, immobile bridge deck

Fig. 5.57. Yscloskey movable lift bridge on LA-46 - damaged operator house

Nearly 80% of the repair costs estimated for the Yscloskey Bridge are for electronics replacement. The total estimated emergency repair cost is $900,000, with only $20,000 for structural repair of barriers and road surface. Included in this estimate is money for cleaning and re-greasing of all pulleys and cables. This is needed because the driving rain and wind of the hurricane stripped oil and grease from these moving parts, causing them to corrode quickly. One of the critical repair items for this and many movable bridges is electronic traffic control gates. Because there are few manufactures able to produce these gates, LADOTD administrators gave this gate shortage problem as an example situation in which repair prioritization was required. The routes the state deemed most important to repair were the first to receive replacement traffic gates.

Bascule Bridge Example: US 11 North Drawbridge
The US 11 North Drawbridge at the north shore of Lake Pontchartrain is one of the few bascule-type movable bridges in Louisiana. Because these bridges require counterweights to move below the bridge deck when the bridge opens, they are not suitable for the low bridges common in the state. Counterweight pits are prone to flooding, which occurred for this bascule bridge. The flooding of the counterweight pits was caused by sump pump mechanical and electrical failures from submersion. Common movable bridge damage occurred on the North Drawbridge from Hurricane Katrina as well, as shown in Figures 5.58-5.59. Navigation lights broke and became detached, and the operator house sustained light damage. The most costly damage to the bridge was abutment erosion and undermining from high flow speeds.

Fig. 5.58. Damaged navigation light on fender of North drawbridge on US11 over Lake Pontchartrain

Fig. 5.59. Broken windows on operator house of North drawbridge on US11 over Lake Pontchartrain

Approximately 25% of the $327,000 estimated for emergency repairs of the drawbridge on US 11 over Lake Pontchartrain will go to replacing damaged electrical and mechanical systems. Because of its height above the lake, the main mechanical and electrical systems for drawbridge operation escaped damage, reducing repair costs substantially. The remaining 75% of the repair budget will go to fixing structural problems such as damaged guardrails and fill around the scoured abutment foundations.

6 RAILROAD PERFORMANCE AND REPAIR

6.1 Performance of Railroads in Alabama

On Sunday, August 28th , 2005 all major carriers suspended rail travel into Louisiana, Mississippi, and Alabama as Hurricane Katrina approached the coast. Six of the nation's seven major freight railroads; the Burlington Northern Santa Fe (BNSF) Railway, CSX Transportation, the Canadian National (CN) Railway, the Kansas City Southern (KCS) Railway, the Norfolk Southern (NS) Railway and the Union Pacific UP) Railway (as well as many short line carriers) suffered damage from Hurricane Katrina (Figure 6.1).
The most common damage was from debris such as trees and barges carried by the storm, covering and damaging railroad tracks, and even knocking over some bridge spans. The tracks had to be inspected, cleaned, and repaired before they could be put back in service. Approximately ten railroad bridges were also damaged, either from impact with objects carried by the storm, or from high wind and storm surge. Several bridges had the track washed off the deck. However, railroad bridges seemed to do better than highway bridges during the hurricane. There were two locations (in Lake Pontchartrain and Biloxi Bay) where the railroad bridge remained standing while a parallel highway bridge lost its

superstructure. This may be due to the fact that railway bridges are designed for much heavier loads.

In addition to damage to tracks and bridges, there was damage to switches and signals and also extensive damage to locomotives, railroad cars, and their contents. After the hurricane, railroad companies enacted the "Force Majeure Clause" to be released from liability for damage caused by circumstances beyond their control. Despite all of the damage, the railroad industry was very proactive before, during and after the hurricane. They expedited the repairs and took steps to minimize their loss and to prevent harm to the community.

The railroads made contact with the state's governors, the U.S. Department of Transportation, and the Federal Emergency Management Administration (FEMA) to help in the recovery. Bottled water, mobile homes, and even recreational vehicles were carried into the region on trains. The freight railroads worked with Amtrak, FEMA, and the Federal Railroad Administration (FRA) to evacuate people out of the area. Those railroad cars that remained in the path of the hurricane had to be hauled away and repaired. All freight cars that were submerged during Hurricane Katrina were required to have a thorough inspection to ensure that the air brake systems and roller bearings were not compromised. All brake valves that were submerged had to be replaced and all cylinders and lines were inspected, cleaned, and repaired according to the Association of American Railroad (AAR) rules. There was no evidence that any dangerous chemicals had leaked from rail tank cars that were overturned in New Orleans by Hurricane Katrina. Mike McDaniel, secretary of the Louisiana Department of Environmental Quality, said, "We have no visual evidence from reconnaissance of leaks or spills." He added that teams are still testing the cars for leaks. He said the state had conducted flyovers of railroad cars using specialized equipment and hadn't discovered any leaks, nor has air-monitoring equipment picked up any evidence of chemicals or toxins in the area from railcars.

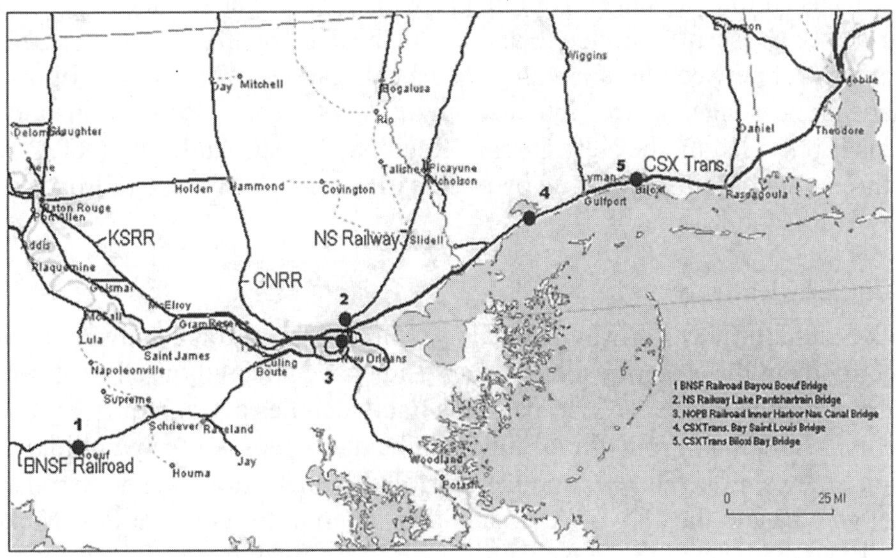

Fig. 6.1. Map of Railroads and Railroad Bridge Damage in Gulf Region.

6.2 Performance of Railroads in Louisiana

New Orleans is a major hub for railroads in the Southern United States. Freight service west of New Orleans is provided by the Union Pacific (UP) and Burlington Northern Santa Fe (BNSF) Railroads (Figure 6.2). Freight service to the east of New Orleans is provided by the Norfolk Southern (NS) and CSX railroads. Freight service to the north of New Orleans is provided by the Canadian National (CN) and Kansas City Southern (KCS) railroads. The area is also served by many short line railroads that include switching yards into the city.

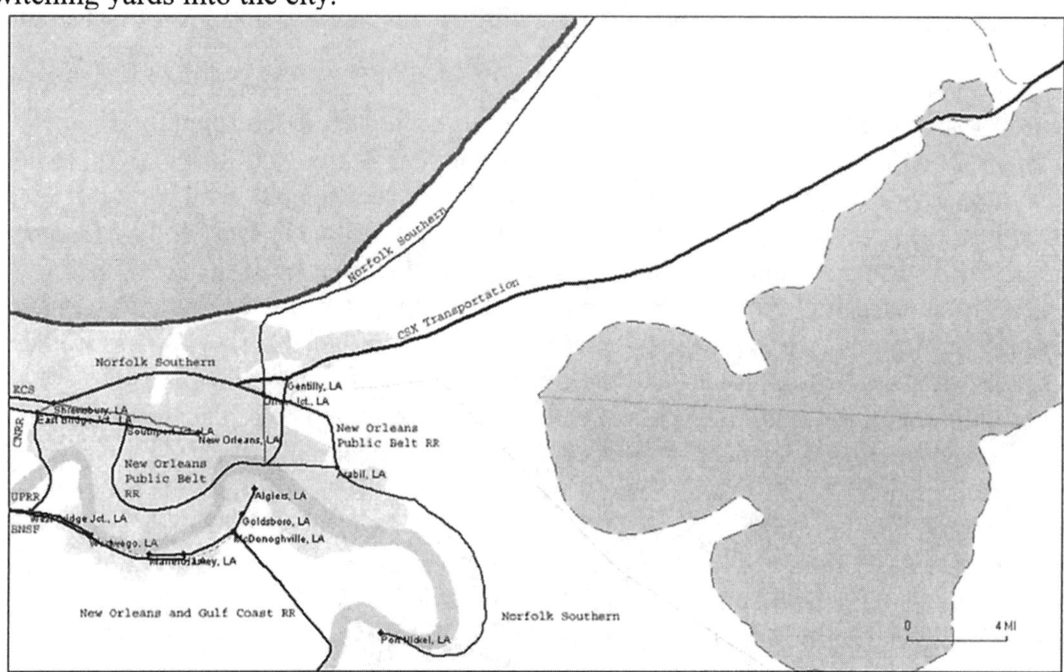

Fig. 6.2. Map of Railroads around New Orleans.

The New Orleans Union Passenger Terminal provides passenger service to and from New York aboard "The Crescent," service to and from Chicago aboard "The City of New Orleans," and travel between Orlando and Los Angeles aboard 'The Sunset Limited." Amtrak manages passenger service into and out of New Orleans. Public transportation in New Orleans is provided by the New Orleans Regional Transit Authority (RTA) which maintains three streetcar lines powered by overhead wires. All of these railroads were impacted by Hurricane Katrina.

Norfolk Southern Railway

The Norfolk Southern (NS) Railway maintains 21,300 miles of track and serves most of the container ports in the eastern United States. They have a heavily used track from Birmingham, Alabama into New Orleans. This track includes a segment from the City of Slidell to Oliver Junction. From Oliver Junction the track goes westerly until it terminates in Shrewsbury. There is another Norfolk Southern track in southern New Orleans that goes east to the City of Arabi and then south until it ends at Port Nickel. All of these tracks are connected to the New Orleans Belt Railway that provides switching to other railroads. It is in this area that the Norfolk Southern Railway sustained most of the damage during Hurricane Katrina. Approximately nine miles of track in the City of New Orleans was underwater or covered by debris after the storm (Figure 6.3).

Following the hurricane, Norfolk Southern inspected the track, removed fallen trees, and replaced ballast, ties, and rails to bring train service back to New Orleans. While these repairs were being made, freight that normally came through New Orleans was rerouted around the city.

Fig. 6.3. Norfolk Southern Railway between Lake Pontchartrain and New Orleans -Sections Were Underwater and Covered with Debris (Courtesy of Norfolk Southern Corp).

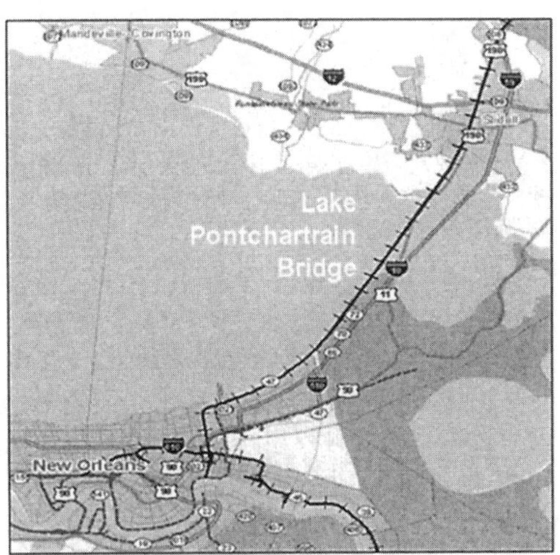

Fig. 6.4 Map of Norfolk Southern Railway across Lake Ponchartrain (Courtesy of Norfolk Southern Corp).

The biggest problem was five miles of track that had been washed off the Lake Pontchartrain Bridge (west of Hwy 11 in Figures 6.4). This 5.8 mile long, reinforced concrete trestle bridge was built between 1983 and 1996 to replace a rotting 30-mile-long timber structure from 1883 (the new bridge is shorter because embankment supports the rail along the shore).

It is supported on 60 foot to 90 foot long, battered 24-inch dia. steel pipe piles with a concrete pile extension. Six piles support a 6 foot wide by 4 foot tall by 17-foot long reinforced concrete bent cap. The piles were driven through the extremely soft lake bottom until they achieved end-bearing in a thin layer of sand. The superstructure is 16' wide by 2'-9" tall by 30' long reinforced concrete double box girder spans. The spans are restrained by a 1-1/2" diameter rebar (a #12 bar) between the bent cap and the soffit in each bay (Figure 6.5). Used rubber belts are placed under the ends of the spans as bearings. These box girders are completely open at each end. Because the bridge is built so close to the water, a bascule span allows barges and pleasure craft access to the shipping channel.

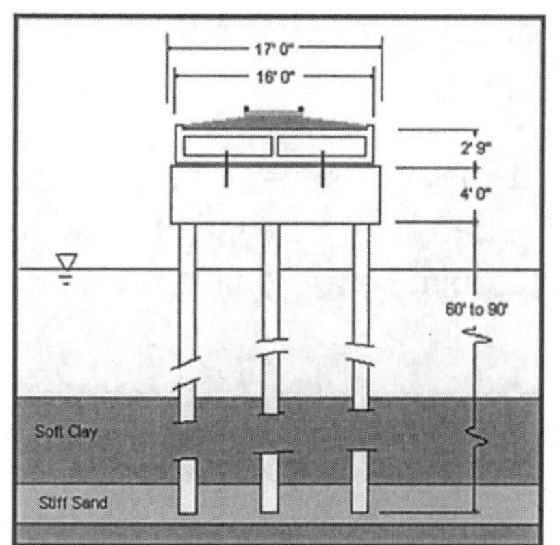

Fig. 6.5. Typical Section of Norfolk Southern Railroad Across Lake Pontchartrain Bridge.

Fig. 6.6. Repairs to the Railroad (Courtesy of Norfolk Southern Corp).

After the hurricane, divers located the unbroken track underwater and nine barge-mounted cranes lifted one mile of track a day back onto the bridge deck (Figure 6.6). Missing ties were replaced and new ballast was poured. Only two spans were lost during the storm and they were replaced by triple-box girder spans that were found in the NS Railway yard. There was very little damage to the bascule span, which was in the closed position during the storm (Figure 6.7). The bridge was rebuilt using a 'Force Contract" where a contractor is hired and paid for labor, equipment, materials, and a predetermined profit. It took 300 people, working around the clock, ten days to reopen the bridge.

Fig. 6.7. Bascule span on Lake Pontchartrain Bridge (Courtesy of Norfolk Southern Corp.).

Within about two weeks, all of the Norfolk Southern Railway was open to train traffic. On October 3rd Norfolk Southern reopened its intermodal terminal in New Orleans. Due to local curfews, the terminal was only open from 8 a.m. to 4 p.m on Monday through

Friday. On the same day, they reopened the Oliver Yard Terminal in New Orleans. This yard serves local industrial customers and interchanges freight with the New Orleans Public Belt Railroad, which serves the Port of New Orleans.

CSX Transportation (CSXT).

Similar to the Norfolk Southern Railway, CSX Transportation has an extensive railroad network across the eastern United States including a line that runs west along the Gulf Coast and terminates at its yard in Gentilly, Louisiana (refer to Figure 6.2). From Gentilly, CSX trains are transferred onto other lines by the New Orleans Public Belt Railroad. There was extensive damage to the 145 miles of track east of New Orleans including considerable damage to the yard at Gentilly (Figure 6.8). Work was immediately begun to repair the Gentilly yard and to move the locomotives and cars out of the area so that they could be used elsewhere in the system.

Fig 6.8. CSX Gentilly Yard after Hurricane Katrina (Courtesy of CSX Transportation.).

CSX hired a prime contractor for the Gentilly yard and for the 145 miles of damaged track east of New Orleans. The contractor sent a 3-mile convoy of vehicles and more than 100 workers into the Gentilly Yards at New Orleans. Most of the water was gone, but it left behind tangles of overturned rail cars, shipping containers, and sections of undermined track. Crews lived in trailers and motor homes for three weeks while they completed repairs. Hundreds of rail cars that were flooded were inspected per the AAR regulations as previously described. Crews used bulldozers with side-mounted booms to pick up overturned cars and locomotives and put them back on the tracks.

The temporary loss of CSX's coastal route is the main reason that rail traffic into the City of New Orleans was severely reduced after the hurricane. Before the storm, about 17,000 cars a month crossed the Huey Long Bridge into the City. After the storm rail traffic was reduced by two thirds.

Burlington Northern Santa Fe Railroad.

The Burlington Northern Santa Fe (BNSF) Railway maintains one of the largest railroad networks in North America with 32,000 miles in 28 western states. The railway is among the world's top transporters of intermodal traffic, moves more grain than any other American railroad, and hauls enough low-sulphur coal to generate about ten percent of the electricity produced in the United States.

The BNSF Railway provides a much-used track between Houston, Texas and New Orleans that sustained considerable damage during Hurricane Katrina. Other than the debris strewn across the mainline to New Orleans, the biggest problem facing work crews was the damage to the Bayou Boeuf Bridge in Morgan City, Louisiana (Figure 6.9). This bridge includes an eight-span steel, through-girder superstructure supported on squat, two-column bents that barely raise the superstructure over the water's surface. Note that two of the spans are actually a swing bridge that can move parallel to the Bayou to allow ships and barges to pass. It was the span east of the swing span that was hit by a barge that broke loose from its mooring during Hurricane Katrina and knocked one end of the span off the bent and into the bayou. The rail and the signal system were broken by the impact (Figure 6.9). Much of the bridge damage during Hurricane Katrina was from barges that bring coal down the Mississippi River.
On August 30th, a tugboat pushed a barge crane into position next to the span to lift it back into place (Figure 6.10). Barge cranes were in great demand after the hurricane. Note that several crews rode the rail to the bridge in trucks to repair the rail and signal. Also note that the much taller US 90 highway bridge just north of the railroad bridge managed to avoid getting hit by the barge. While the bridge was being repaired, BNSF rerouted freight through St. Louis, Missouri, Chicago, Illinois, and Memphis, Tennessee. The BNSF main line was reopened on September 1st after repairs were completed to the bridge over Bayou Boeuf in Morgan City, Louisiana.

Fig. 6.9. Shifted Spans on the Bayou Boeuf in Morgan City, LA (Courtesy BNSF Railway).

Fig. 6.10. Repairs to Bayou Boeuf Bridge in Morgan City, LA (Courtesy BNSF Railway).

Canadian National Railway.

The Canadian National Railway owns hundreds of miles of track and a dozen facilities along the Gulf, including a freight yard in New Orleans (which was on higher ground). The segment that runs across the LaBranche Wetlands in St. Charles Parish, was damaged by Katrina, (and again by Rita). Storm surge washed out the roadbed beneath the tracks where they cross the wetlands, but repairs were completed in a few days and there was no bridge damage. The New Orleans service is by way of a line from Baton Rouge, with the direct line between Hammond, LA and New Orleans (Figure 6.11), and this line was reopened a few days after the hurricane.

During the weekend before Katrina hit New Orleans, crews moved locomotives and other equipment out of the area and then halted all traffic. The storm took down the intermodal facility near New Orleans, where containers are transferred between train cars and trucks. Grain shipments were slowed down due to disruptions to employees, power supplies, and communications lines. Three months after Katrina, things were still slow, but in the long run, the hurricane may improve the economy of the region, especially for the CN Railway. They are the largest hauler of forest products in North America. They're looking at putting in new infrastructure for transferring lumber from railcar to truck somewhere in the New Orleans area. They're also exploring potential opportunities for removing the debris out of New Orleans.

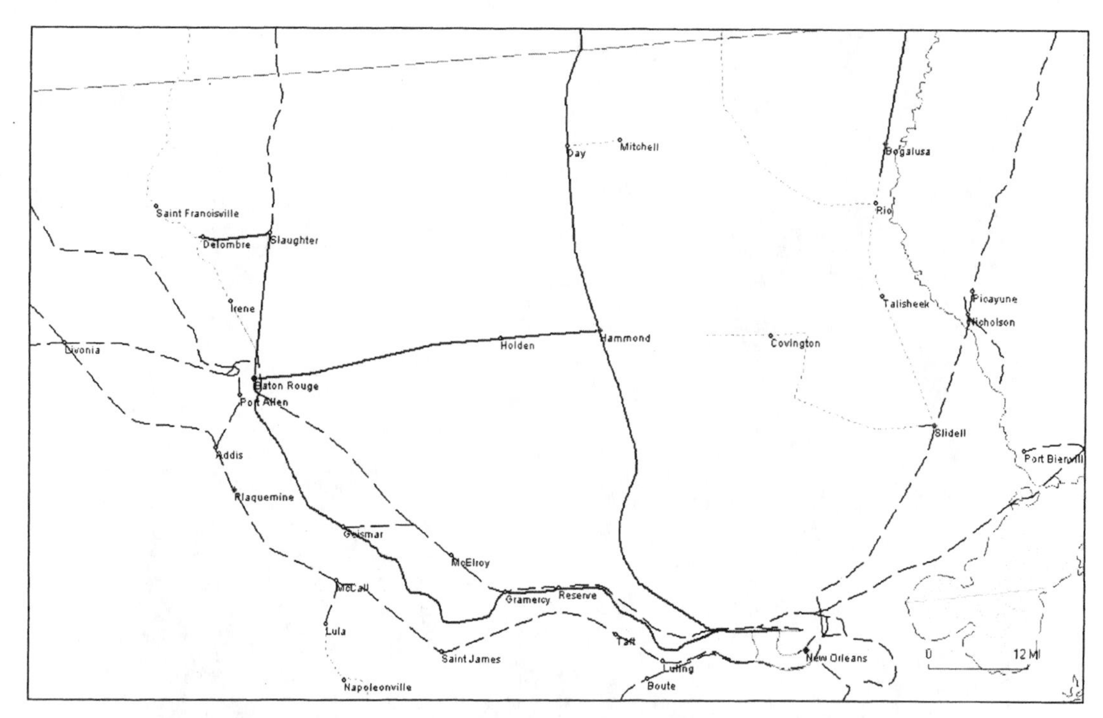

Fig. 6.11. Map showing Canadian National Railway tracks in Louisiana.

The Union Pacific and the Kansas City Southern Railways.

The Union Pacific (UP) Railway reported that its lines to the west of New Orleans appear to have been spared major structural damage. Inspectors have declared that the Huey P. Long Bridge is safe for rail operations. UP uses that bridge to exchange an average of 14 trains a day with CSXT and NS.

The Kansas City Southern (KCS) Railway restored service along its Meridian Speedway between Meridian, MS and Shreveport, LA on August 31. Service along remaining lines was restored on September 2, 2005.

New Orleans Public Belt Railroad.

About 1/3 of the nation's east-west rail freight crosses the Mississippi River on the Huey P. Long Bridge, but damage to railroad tracks by Hurricane Katrina severely reduced train travel into New Orleans. Most train traffic was rerouted to Memphis, Tennessee and St. Louis, Missouri. There was damage to approximately 10 miles of New Orleans Public Belt Railroad track between Poland Avenue and Gentilly. Repairs took six weeks and cost about $8 million. New Orleans Public Belt, which provides rail interchange service to the Port of New Orleans, was repaired about six weeks after the storm.

New Orleans and Gulf Coast Railroad.

The New Orleans Gulf Coast (NOGC) Railroad maintains tracks south of the City and west of the Mississippi River. Northern operations were resumed almost immediately after the storm but southern operations were suspended from mileposts 13 to 24 due to storm damage that removed ballast and covered the right-of-way with debris. The railroad was back in service a month after the storm.

Amtrak.

"The City of New Orleans" was operating (as of November 25, 2005) between Memphis and Chicago instead of between New Orleans and Chicago as a result of the hurricane. "The Crescent" was operating between Atlanta and New York instead of between New Orleans and New York beginning on October 9, 2005. Service that had been provided to New Orleans on a three-days-weekly schedule by the Sunset Limited (trains 1 & 2) to Orlando and Los Angeles has not been restored as of March, 2006. The westbound "Sunset Limited" originates in San Antonio, Texas instead of from Orlando, Florida. Service to the east, via Biloxi and Mobile, was not be possible until 2006 because of all of the damage to tracks, bridges, and infrastructure owned by CSX Transportation.

New Orleans Regional Transit Authority.

As previously mentioned, there are three street car tracks maintained by the RTA: the recently completed Canal Street line, the 20 year old Riverside line, and the 170 year-old St. Charles line. All 24 cars on the Canal Street line and six of seven cars on the shorter Riverfront line were damaged by the flooding that occurred followed the hurricane. Cars on the old St Charles line were fine, but the power system was destroyed and had to be rebuilt. They had parked the old cars in the Uptown barn, which escaped the flooding. They had parked the new cars at the Canal Street barn that was supposed to be the main evacuation point, but it had over five feet of water. The water was apparently extremely corrosive and destroyed the undercarriage, the wiring, the seats, the paneling, and the exterior. It took 142 days to build each new car and will take about as long to repair them, at a cost of $1 million per car. The price tag for repairing the power lines on the St. Charles Street is lower, but has to wait until electricity has returned. The St. Charles cars are able to run on the other lines, but because of their historic designation, it is not allowed.

6.3 Performance of Railroad Bridges in Mississippi.

Almost all of the rail damage in Mississippi was to the CSX Transportation Corporation whose track runs less than a mile from the Gulf Coast. There was some damage to a few of the short line railroads in the state, but the bulk of the rail damage was to the CSXT. This damage diminished east of Biloxi Bay and there was very little railroad damage in Alabama.

CSX Transportation (CSXT).

The Mississippi coastline was severely damaged and consequently, almost all of the CSX railway along the Mississippi Gulf Coast was damaged as well. CSXT has little it can do but continually repair their tracks and bridges as hurricanes are a continual threat along the Gulf. In fact, the railroad embankment (about 26 feet above sea level) protected residents living north of the track from flooding and storm surge. Consequently, CSXT and US90 took the brunt of the storms fury. Figure 6.12 shows the typical damage to the CSXT railroad after Hurricane Katrina in Waveland on the western side of Bay Saint Louis.

Fig. 6.12.Damaged CSXT Track (Courtesy of CSX Transportation).

Bay Saint Louis Bridge

The Bay Saint Louis Bridge was a two-mile long structure composed of a reinforced concrete box girder simply-supported spans supported on alternating three and four column bents. Note the very small shear keys that don't offer much lateral resistance for the superstructure (Figure 6.13).

Fig. 6.13. Undamaged Portion of CSXT Bay Saint Louis Bridge(Courtesy of Prof. Yim/OSU).

During the hurricane, most of the rail, along with many of the spans were pushed off the bridge, probably as a result of high winds and storm surge (Figures 6.14-6.15).

Fig. 6.14. Damage Near Swing Span of CSXT Track Over Bay St. Louis Bridge

Fig. 6.15. Damaged Span of CSXT Track in Bay St. Louis (Courtesy of Prof Yim/OSU).

Repairs to the Bay Saint Louis Bridge were completed on March 1st under a 'Force Account' contract. Damage to this bridge is just part of $250 million worth of CSX damage from Katrina, according to Chief Operating Officer Tony Ingram. Because the piers survived Katrina intact, with only a little work, they were ready for a new deck and tracks. Starting at Bay St. Louis, the contractor placed four, 60-foot-long, precast I-girders across the spans. Formwork was placed between the girders, reinforcing steel was laid out, and a concrete deck was cast. After the deck was finished and cured, the railroad ties and rail were set in place and covered with ballast. The contract was to span 110 of the 160 gaps and fix the swing span that lets barge and boat traffic through the middle of the bridge. Starting in Pass Christian, another company had a contract for the remaining 50 spans. Supporting the crew of 90 workers are eight barges, two tug boats, two high-speed skiffs, five cranes from 100 to 230 tons, as well as trucks, forklifts, generators, arc welders and compressors.

Biloxi Bay Bridge

The CSXT Biloxi Bay Bridge was another two-mile long bridge. However, unlike the Bay St. Louis spans, the Biloxi Bay Bridge is composed of four simply-supported precast I girders with a cast-in-place deck (along with ties, rails, and ballast). It is supported on pile caps with six battered precast piles. Fifteen-inch high shear keys restrain the superstructure from lateral movement and 1.25 inch diameter through-bolts provide lateral stability to the girders.

During the hurricane, the rails and ballast were thrown into the Bay but the superstructure remained intact (Figures 6.15 to 6.16). If the Bay Saint Louis Bridge with its small shear keys and loss of superstructure was a failure, than the Biloxi Bay Bridge was a great success, especially since the adjacent US 90 bridge was severely damaged. Figure 6.17 shows repairs taking place on the Biloxi Bridge in early November. Ties and rail were laid on the deck in preparation for new ballast to be placed between the barriers.

After Hurricane Katrina, approximately 100 miles of track from New Orleans to Pascagoula, Mississippi were closed to train traffic with some tracks ripped from the ground and others covered by debris, included houses and barges (Figure 6.18).

Fig. 6.15. Damage to Biloxi Bridge - Missing Rail, Ties, and Ballast (courtesy of Prof. Robertson/U of H).

Fig. 6.16. Debris on Biloxi Bridge after Storm (courtesy of Prof. Robertson/U of H).

Fig. 6.17. New CSXT Tracks being placed on Biloxi Bridge.

Fig. 6.18. Closure Due to Debris (Courtesy of CSX Transportation).

Port Bienville Railroad (PBR)

Port Bienville Railroad closed 80% of their 14.5 miles of track after Katrina. The track had been washed out, 350 cars had been submerged above the wheels, 122 cars were derailed, 2 locomotives were submerged above the frames, and 22 employees lost their homes (the railroad purchased and set up 12 mobile homes) and their connection with CSX was destroyed. PBR customers are reporting 6,000 jobs affected by Katrina, which was losing $250,000 per day because they are shipping by truck, and the PBR was losing $85,000 each day while their connection is down. Total damage estimates, including temporary housing is $7,150,000. Removing the shrimp boats from the tracks was made more difficult because the U.S. Coast Guard (fearing diesel fuel in the boats' tanks might cause an environmental hazard) halted removal of the boats. Repair crews had to pump out all of the fuel before they could move the boats.

7 ROADWAY PERFORMANCE AND REPAIR

7.1 Mississippi Roadways

The major impedance to roadway traffic following Hurricane Katrina was the extensive amount of debris from the storm. Debris removal in Mississippi has been estimated to cost $100 million for state highways and $35 million for county routes. Sixteen hours after the event, all MDOT maintained roadways were opened to emergency traffic, and most were opened to the general public within three days, with the exception of US-90.

Scour to highways 603 and 43 near Waveland were reported by inspectors and quickly filled and repaired. However, the 26-mile stretch of US90 between Pass Christian and Biloxi suffered severe damage. Despite the 14' seawall along the coast, the storm surge rose to levels of 25'-30' along the route. Following the storm, 4'-5' of sand covered the roadway and took a week to be removed. This revealed the underlying damage to the roadway, where asphalt pealing from the sand base occurred as well as break-up of sections due to differential settlement and washout of the base (Figures 7.1-7.2). Two months after the event, two lanes were opened, allowing one east- and west-bound lane. Damage to the infrastructure along US90 also has impacted its availability and restoration. This includes the need to repair and clear storm water drains, replace damaged box culverts, in addition to repairing and replacing sections of the washed out roadway, as seen in Figures 7.3-7.4. As of November 7, 2005, traffic along the highway was very light, as many of the surrounding homes and businesses had suffered severe damage. The 4-lane highway was fully opened on December 17, 2005, though the bridges at either end of the 26 mile stretch will take significantly longer to replace.

Fig. 7.1. Debris to 26 mile section of US-90 between Pass Christian and Biloxi

Fig. 7.2. Damage to 26 mile section of US-90 between Pass Christian and Biloxi

Fig. 7.3. Repair of US-90 roadway and stormwater system including clearing of drains

Fig. 7.4. Repair of US-90 roadway and stormwater system including replacement of culverts

7.2 Performance of Roads in Louisiana

The major impedance to roadway traffic following Hurricane Katrina was the extensive amount of debris from the storm and the damage that occurred at key bridges in the area. As of November 2005 FEMA estimated that more than 21.1 million cubic yards of debris had been removed in Louisiana. Most debris removal costs incurred by state and local governments will be reimbursed 100 percent by FEMA until Jan. 15, 2006. As of March 3, 2006, FEMA has obligated more than $65 million in federal dollars for debris removal in Louisiana as a result of Hurricane Katrina.

Fig. 7.5. West End Boulevard, New Orleans

Fig. 7.6. Debris Piling Up on West End Boulevard, New Orleans

7.3 Performance of Roads in Alabama

Hurricane Katrina caused at least $20 million dollars worth of damage to Alabama's roads (Figure 7.7). This is lower than the storm damage in Louisiana and Mississippi and it's also lower than the damage from 2004's Hurricane Ivan, which cost the state an estimated $27 million. Mobile sits slightly inland on the Mobile River, on the eastern side of the storm's eye. After the hurricane, many streets, roads, and highways were underwater. There was congestion and delays along main routes across southwest Alabama especially in urban areas and along major intersections and interchanges. State and city agencies reopened most of the roads within a day or two of the storm. ADOC State Inmate labor assisted state and local government officials in clearing roads.

The most significant road damage included the following:

- Downtown Mobile was under several feet of water after the hurricane.
- The Bankhead Tunnel was closed during the height of the storm but reopened following Katrina's departure.
- The Wallace Tunnel suffered minor flooding and remained open to at least one lane of traffic per direction in the wake of the departing storm with pumping equipment in the other lanes to expel floodwaters from the ventilation and drainage shafts under the driving surface.
- ALDOT crews quickly repaired travel lanes along Battleship Causeway eroded by the storm surge. All lanes were reopened by September 2, 2005. The Exit 30 eastbound on-ramp to Interstate 10 remains closed as five concrete bridge spans were destroyed.
- Dauphin Island Bridge (Alabama 193) suffered tremendous damage. After the storm, the causeway linking the barrier island to South Mobile County had one lane open with a pilot car. The bridge reopened to two lanes of traffic on September 2, 2005. Crews repaired washouts and shoulder damage.
- The fiber optic cables to the I-10 traffic information signs sustained major damage. Extensive repairs are necessary and this traffic information system will be out of service for an extended period until repairs are made.
- Fort Morgan Road is fully open to traffic with debris cleanup and sand removal continuing in some places.
- Immediately after the storm, US90/98 causeway in Baldwin County was closed due to water over the bridge (Figure 7.8).
- Immediately after the storm, I-10 to Mississippi was open to emergency vehicles only.

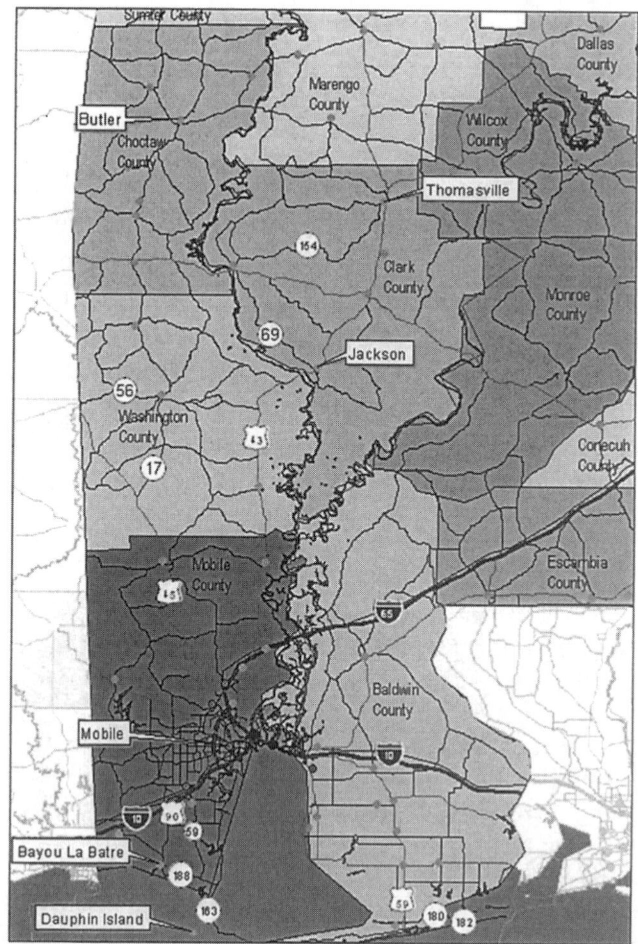

Fig. 7.7. Locations of roads in southern Alabama where storm damage occurred.

Fig 7.8. Approach roadway damage at Tensaw River on US-90/98 in Mobile, AL

Table 7.1 provides a description of the extent of roadway damage immediately after the hurricane. It also lists roads that were closed in anticipation of the storm. Table 7.2 describes the roads that remained closed three months after the hurricane.

Table 7-1 Road and Highway Closures as of August 30th, 2005

Route	Status	County	Location	Reason	Posted
AL 180	Closed	Baldwin	From Fort Morgan to MP 6	Debris in roadway	8/30/2005
AL 59	Closed	Baldwin	From SR 180 South to Beach Rd	Debris over roadway	8/30/2005
SR-69	Advisory	Marengo	M.P. 44.7 near Clarke/Marengo County Line	Trees and power lines across roadway.	8/30/2005
SR 171	Advisory	Tuscaloosa	Between MP 3 and MP 12	Roadway open, but power lines down	8/30/2005
SR 7 (U.S. 11)	Closed	Tuscaloosa	Near B F Goodrich Tire Plant (appox. MP 71)	Roadway blocked with tree/power lines down	8/30/2005
SR 69 North	Closed	Tuscaloosa	Between MP 168 and MP 177, Co. Rd. 38 to Co. Rd. 46	Roadway closed - trees down with power lines	8/30/2005
SR 56	Advisory	Washington	From Mississippi state line to U.S. 43	Trees down along roadway at a number of locations. Drive with caution	8/30/2005
I-10 EB and WB	Advisory	Mobile	I-10 at Wallace Tunnels in Mobile	Proceed through I-10 Wallace Tunnels with caution. One lane open in each direction. Hazardous cargo is prohibited through Wallace Tunnels. Hazardous cargo detour is along SR 59 and I-65 due to closing of Cochrane Bridge.	8/30/2005
SR-17	Advisory	Choctaw	From Washington/Choctaw County line to Butler.	Trees and power lines down at a number of locations along roadway. Drive with caution	8/29/2005
SR-5	Advisory	Wilcox	Pine Hill	Signal Out	8/29/2005
SR-17	Advisory	Washington	From U.S. 45 to Choctaw County Line	Trees and power lines down along roadway at a number of locations. Traffic signals out in Chatom at SR-56. Drive with caution.	8/29/2005
US-84	Advisory	Choctaw	From Mississippi State line to Grove Hill	Trees and power lines down at a number of locations along roadway. Drive with caution.	8/29/2005
SR-154	Closed	Clarke	From 4 mi. East of SR 69 (MP 4)to 17 miles West of US 43 (MP 12)	Trees in Roadway.	8/29/2005
US 98	Closed	Mobile	Cochrane-Africatown USA Bridge	ALDOT is currently inspecting the bridge for possible structural damage.	8/29/2005
U.S. 45	Advisory	Washington	From SR-17 to Mississippi state line.	Trees down along roadway at a number of locations. Drive with caution	8/29/2005
I-10 & US 90	Advisory	Mobile	Approaching Mississippi Stateline Westbound	Extensive damages to I-10 and US 90 in MS due to Hurricane Katrina. Both Roadways in MS Closed. Seek alternate routes.	8/29/2005
SR-69	Advisory	Clarke	From U.S. 43 at Jackson north to Marengo County line.	Trees down along roadway at a number of locations. Drive with caution	8/29/2005
All Routes	Open	Conecuh		All routes open.	8/29/2005
All Routes	Open	Escambia		All routes open	8/29/2005
SR-17	Advisory	Sumter	From U.S. 11 to Pickens County line.	Trees down along roadway at a number of locations. Drive with caution.	8/29/2005
Welcome Center I-59/20	Closed	Sumter	I-59/20 @ Miss. State Line	Closed due to Power Outage	8/29/2005
US 98	Open	Baldwin	AT AL 104	Road has been reopened.	8/29/2005
US 90/98	Closed	Baldwin	Causeway	Water over the road.	8/29/2005
AL 182	Closed	Baldwin	Gulf Shores	Closed due to water and debris over roadway.	8/28/2005
AL 193	Advisory	Mobile	Between AL 188 and Dauphin	Roadway is open with pilot car	8/28/2005

			Island	assistance	

Table 7.2. Road and Highway Closures as of November 20th, 2005

Route	Status	County	Location	Reason	Posted
Mobile Bay	Closed	Mobile	Dauphin Island/Fort Morgan Ferry	Closed until further notice due to hurricane damage.	10/27/2005
SR-154	Advisory	Clarke	From 4 mi. East of SR 69 (MP 4)to Thomasville	Debris along side of roadway. Proceed with caution.	8/29/2005
SR-17	Advisory	Choctaw	From Washington/Choctaw County line to Butler.	Trees down at a number of locations along roadway. Drive with caution	8/29/2005
SR-17	Advisory	Washington	From U.S. 45 to Choctaw County Line	Trees down along roadway at a number of locations. Drive with caution.	8/29/2005
SR-182	Advisory	Baldwin	From Little Lagoon Pass westward	Debris cleanup on roadway is complete. Open to local traffic. Check with local authorities regarding access to area.	8/28/2005
SR-193	Advisory	Mobile	BETWEEN AL 188 AND DAUPHIN ISLAND	The roadway was heavily damaged in recent hurricanes. It is now opened to two way traffic. Expect delays during the day due to on going repair work.	8/28/2005
SR-56	Advisory	Washington	From Mississippi state line to U.S. 43	Trees down along roadway at a number of locations. Drive with caution	8/30/2005
SR-69	Advisory	Clarke	From U.S. 43 at Jackson north to Marengo County line.	Trees down along roadway at a number of locations. Drive with caution	8/29/2005
US 90	Advisory	Mobile	Cochrane-Africatown USA Bridge	The Cochrane Bridge is open with one lane in each direction. Overweight vehicles are prohibited from crossing the bridge.	8/29/2005

8 REROUTING AND TRAFFIC DEMANDS

8.1 *Mississippi*

Damage to bridges and roadways along the Mississippi Coast affected the routes available for emergency response, transport of goods and supplies for relief, and travel to residences. The initial emphasis in relieving traffic demands was on getting I-10 cleared as a part of the national defense network, in order to allow through traffic to continue followed by addressing local traffic issues. Clearing of US49 and US59 became priorities for allowing transport of supplies down to the Coast. With the East-West route of US90 and the two bridges at Bay St. Louis and at Biloxi-Ocean Springs completely damaged, I-10 was the only primary route for travel along the Coast. Natural detours from the Coast to I-10 became apparent, such as the use of highway 609 or highway 43/603, though they were not officially posted. Traffic counts performed in 2005 after

Katrina indicated an increased demand on I-10 from the pre-Katrina count performed in 2004, as shown in Figure 8.1.

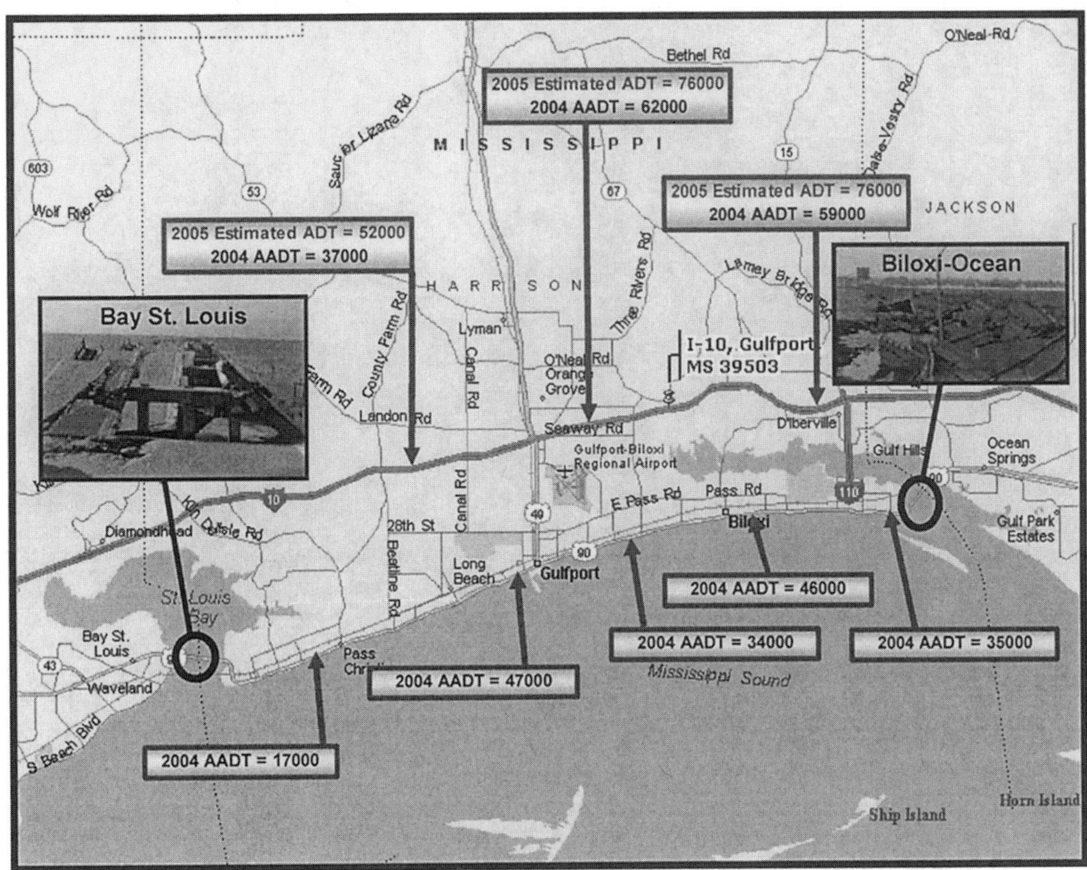

Fig 8.1. Increased average daily traffic count on I-10 after Hurricane Katrina.

Damage to the Biloxi-Ocean Springs Bridge and reduced capacity of I-110 due to the Biloxi Back Bay Bridge damage led to limited access to the Biloxi peninsula and severe traffic on I-110. A detour around the damaged Biloxi-Ocean Springs Bridge was to travel up I-110 (further facilitated once repairs were complete), across I-10, and back down highway 609, as seen in Figure 8.2. In order to accommodate the increased demands placed on I-10 between 609 and I-110, MDOT added 1 lane of traffic after approximately one month of night paving. In addition, a traffic signal was added at the intersection of I-10 and 609 to prevent back-up of traffic on the I-10 off ramp from travelers detouring around the damaged bridge. Ongoing traffic study is being performed to retime the traffic lights to accommodate the increased traffic on 609.

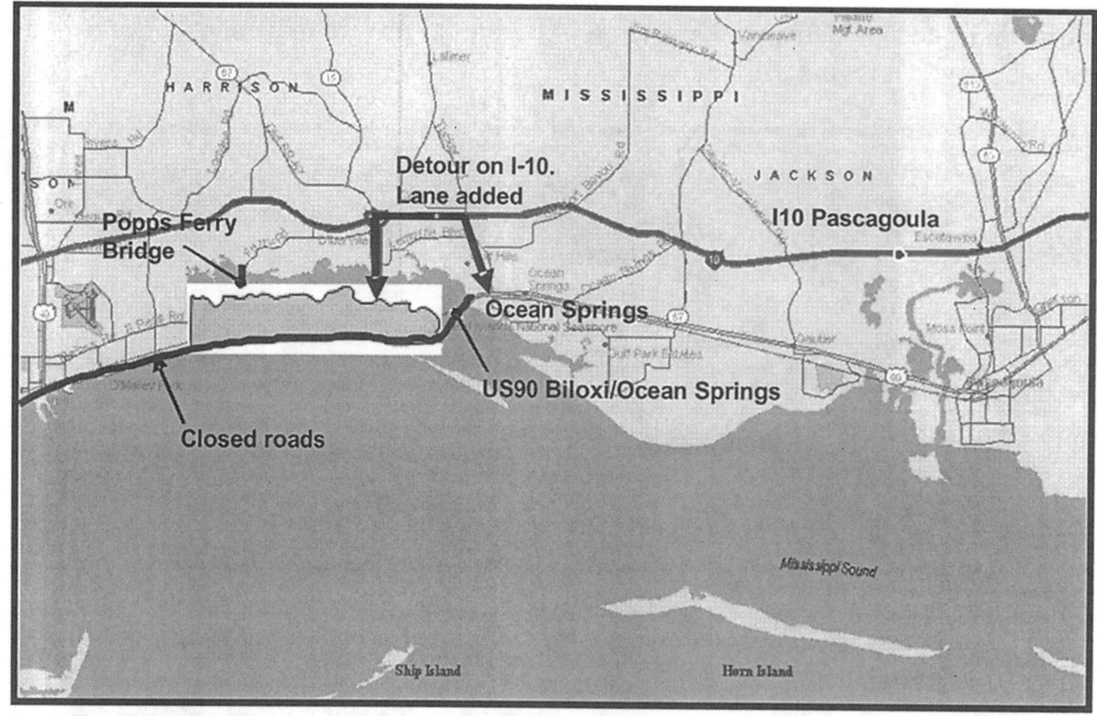

Fig. 8.2. Detour around Biloxi-Ocean Springs Bridge and increased capacity of I-10.

8.2 Rerouting and Traffic Demands in Louisiana

The following key bridges were damaged severely enough to impede roadway traffic and therefore cause traffic closures after Hurricane Katrina.

- I-10 Twin Span Bridge over Lake Pontchartrain
- US-11 Bridge & Pontchartrain Causeway
- Chef Mentur Pass Bridge, US 90
- Rigolets Pass Bridge

Overall damage to Louisiana roads was minimal and mostly based on washouts. There was some erosion observed along Lakeshore Drive in New Orleans, Louisiana. Hurricane Katrina effects on the roadways will be seen long-term by the surge in population in other Louisiana cities such as Baton Rouge, Lafayette, Houma, and Hammond. With this increase population comes an increase in traffic volume that will undoubtedly decrease the life span of roadways that were not designed for this type of traffic volume.

Fig. 8.3. Lakeshore Drive, New Orleans (11-10-2005)

Fig. 8.4. Damage to Lakeshore Drive, New Orleans

8.3 Rerouting and Traffic Demands in Alabama

As was previously stated, after the hurricane, many streets, roads, and highways were underwater. There was congestion and delays along main routes across southwest Alabama especially in urban areas and along major intersections and interchanges. Authorities in Mobile reopened the Cochrane-Africatown USA Bridge to traffic Tuesday, August 30th after inspectors determined it had suffered only minor to moderate damage after being struck by a loose oil rig. As a precaution, however, the four lanes of the bridge were being reduced to one lane in each direction, said Alabama Department of Transportation spokesman Tony Harris. The bridge, part of U.S. 98, was the detour route for vehicles hauling hazardous materials not allowed in downtown tunnels on Interstate 10. The closure of the bridge forced trucks hauling hazardous material to make a 70-mile detour. The Bankhead tunnel, which takes U.S. 98 under the Mobile River, also was closed by water covering its entrances. The Interstate 10 tunnels in downtown Mobile were spared, but only one lane in each direction was available because of pumping operations to keep the tunnels dry, according to the Alabama DOT. Pavement in the tunnels was wet, but there was no standing water. I-10 was passable through Alabama, but only to the Mississippi state line, Harris said. Aerial footage of Dauphin Island off the coast showed flooding, but most of the structures appeared to be mostly intact.
The eastbound US 90 to eastbound I-10 interchange over Mobile Bay suffered structural damage. The closure of this bridge did not appear to have a significant impact on traffic since the US 90 and I-10 run parallel to each other, alternate routes are available, and other on-ramps are available within three miles of the closure.

9 IMPACT ON NEW DESIGN

The nature of the bridge damage during Hurricane Katrina has highlighted some of the vulnerabilities of existing bridges. Discussions pertaining to needed changes in new designs have resulted in a number of proposed changes. According to Paul Fossier Jr., an assistant state bridge design engineer for LaDOTD, FHWA is now requiring a storm surge analysis to be conducted for all new bridges over coastal waterways. The analysis should provide surge loads which should be explicitly considered in design.

One method for handling storm surge is to raise the elevation of bridges above the anticipated storm surge level thus eliminating the loads altogether. Unfortunately, not all sections of the bridge are capable of being elevated. To bring the bridge down to the level of roadways, bridge sections located at the approaches will have to be placed below the anticipated surge levels. Therefore, other details are being considered to handle the loads. Air vents in the bridge decks and diaphragms are proposed to allow for air dissipation and thus a reduction in buoyancy forces. Placing some type of vertical restraint devices, such as steel cables seen in Figure 9.1, is also being discussed. Additionally, measures for protecting reinforcing steel from corrosion to maintain a high level structural integrity are being implemented.

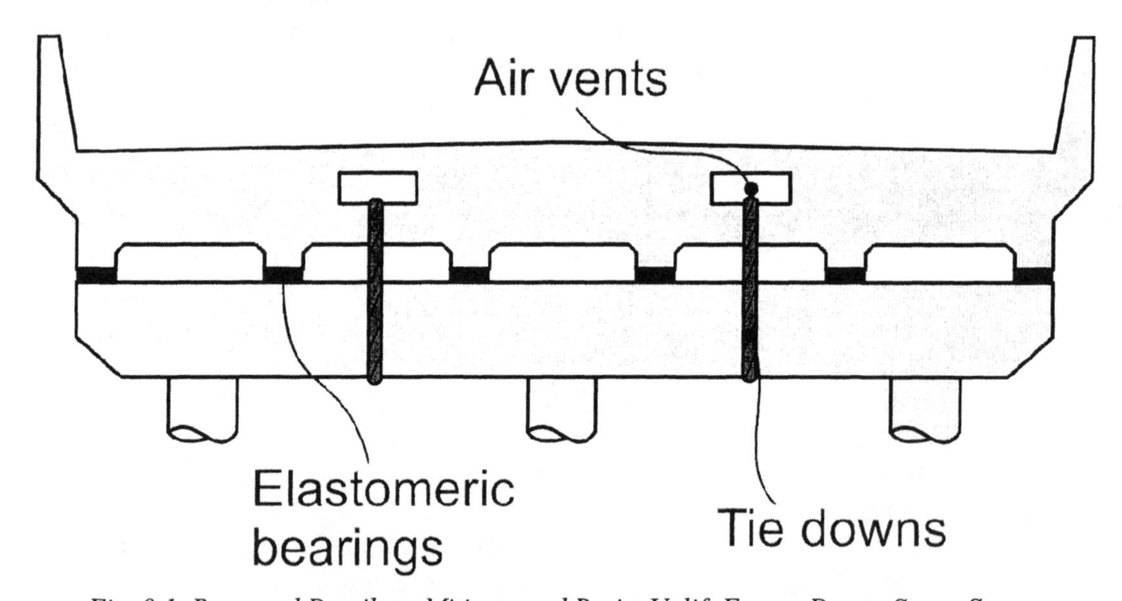

Fig. 9.1. Proposed Details to Mitigate and Resist Uplift Forces Due to Storm Surge.

There were few immediate changes to bridge design practice in Alabama as a result of Hurricane Katrina. Fred Conway, the chief bridge engineer for the Alabama DOT expressed frustration that there was no AASHTO criteria to design bridges for storm surge. Until one is developed, Alabama DOT will more securely anchor girders to the bent cap to resist storm surge (Figure 9.2).

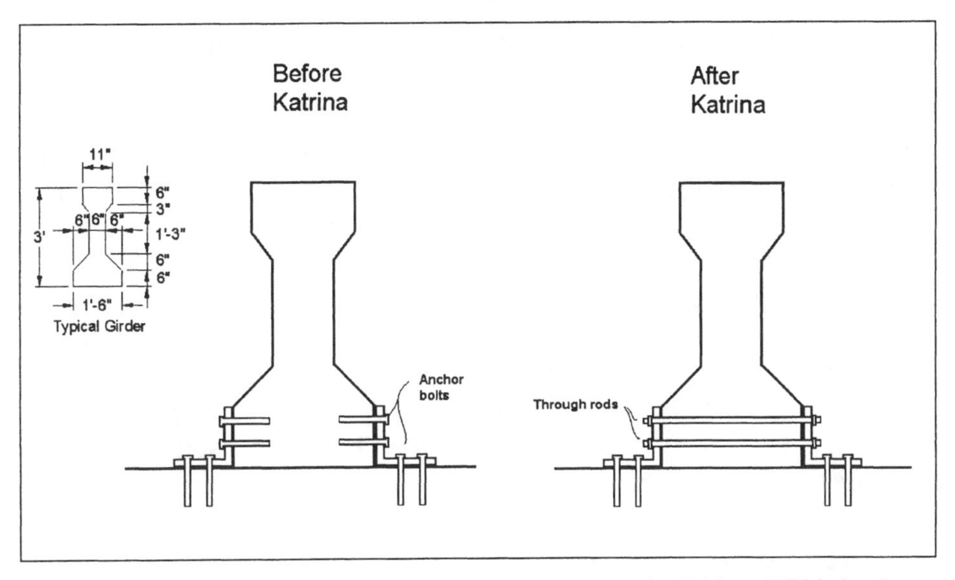

Figure 9.2. Modification for the replacement spans on the I-10 to US90 On-Ramp.

The FHWA sponsored a meeting of state DOT's at risk from hurricanes in December 2005 to discuss changes to design criteria based on lessons learned after Katrina. This meeting helped to set goals for future research. Scientists and engineers have commented on the similarities between storm surge and tsunami. A $1.25 million research contract (sponsored by the University of Hawaii in 2006) at the NEES tsunami wave basin at Oregon State University may help in the development of bridge design criteria for tsunami and storm surge.

10 CONCLUSIONS AND LESSONS LEARNED

The storm surge and large winds from Hurricane Katrina resulted in significant damage to bridges and the transportation network in the Gulf Coast region. Approximately 45 bridges had moderate to significant damage, at an estimated overall cost (including both emergency repairs and replacements) of over 1 billion dollars (*note, this estimate will change as bridge replacement costs change over the coming years*). Emergency preparedness and response plans that were implemented by the departments of transportation of Mississippi, Louisiana, and Alabama were effective in the quick evacuation of residents through the hurricane evacuation routes, providing traffic control, and facilitating communication and coordination between agencies. The hurricane also resulted in major disruption to the railroad system in the southeastern US. As the storm approached, all major carriers suspended rail travel into the storm region. The damage to the railroad system was in the form of debris (trees and barges) being deposited on the tracks, as well as damage to the tracks themselves. Several rail bridges had damage due to storm surge, although, in general, they appeared to perform better than highway bridges.

Innovative repair strategies were used to get bridge functionality restored as rapidly as possible. In addition, incentive plans proved effective for getting the most critical bridges open as quickly as possible. The damage patterns from the hurricane show that bridges close to the coast were most vulnerable, due to storm surge forces. However, it appears that simple measures, such as transverse shear keys and air vents, may be effective in limiting the damage to bridges subjected to storm surge forces. This is an area where additional research by the DOT's is needed. Below, a statement from J. Richar Capka, Acting Administrator, Federal Highway Administration to the US Legislature highlights the steps being taken from the FHWA to ensure better performance of transportation facilities in future storms:

> *We have begun a review of existing bridges that might be impacted by storm surge conditions in the future. Before we can identify suitable retrofits for existing bridges of the types damaged during recent hurricanes, we must improve our understanding of, and ability to quantify, the lateral/transverse and uplift forces that result from floods and storm surges. Accordingly, we have initiated research at the Turner- Fairbank Highway Research Center to aid our understanding in this area. With respect to the design of new bridges, FHWA has developed a policy that defines a flood frequency approach for the hydraulic analysis and design of coastal bridges. We also are reviewing the problem of loose barges impacting bridges during storm conditions.*
>
> *Contraflow is an emerging traffic operations area that requires close coordination of all levels of government. We recognize the challenges of evacuation and contraflow and the need for more attention to these areas in the future. As we did after Hurricane Ivan in 2004, we will analyze the events of Hurricane Katrina for lessons learned that can be applied to future situations. We also will continue to work with other Federal agencies to determine where transportation assets and systems can continue to contribute to evacuation planning and execution. FHWA will assist the Office of the Secretary of Transportation and the Department of Homeland Security in developing the Catastrophic Hurricane Evacuation Plans Report to Congress as mandated in the Safe, Accountable, Flexible, Efficient Transportation Equity Act: A Legacy for Users (SAFETEA-LU).*

The complete text of J. Richard Capka's speech before the US House of Representatives can be obtained at: www.house.gov/transportation/highway/10-20-05/capka.pdf

Index